HALF GONE

Oil, Gas, Hot Air and the Global Energy Crisis

Jeremy Leggett

Portobello
BOOKS

Published by Portobello Books Ltd 2005

Portobello Books Ltd
Eardley House
4 Uxbridge Street
Notting Hill Gate
London
W8 7SY

A CIP catalogue record is available from the British Library

9 8 7 6 5 4 3 2 1

ISBN 1 84627 004 9

Designed by Patty Rennie

Typeset in Janson by Avon DataSet Ltd, Bidford on Avon,
Warwickshire

Printed in England by MPG Books Ltd, Bodmin, Cornwall

www.portobellobooks.com

Contents

Acknowledgements

Though the final text is entirely my own responsibility, I have benefited from considerable help and editorial advice along the way. Those who have commented helpfully on my evolving drafts include Colin Campbell, Colin Hines, Chris Skrebowski, Henner Ehringhaus, Michael Smith, Richard Hardman, Paul Hohnen, Richard Holton, Roger Booth and Stephan Schmidheiny. Other reviewers of crucial sections of the book, from the oil industry, must remain anonymous. Invaluable research was compiled for me over a period of six months by Harriet Williams. My agent Tif Loehnis, publisher Philip Gwyn Jones and editor Jane Hindle have been a constant source of helpful suggestions and encouragement. My family and colleagues at solarcentury have tolerated my workaholic habits along the way, often indeed encouraging them by collecting articles of relevance as the drumbeat of concern about oil depletion and global warming has steadily built.

London, August 2005

The Story of the Blue Pearl

Once upon a time there was a planet on which life evolved into such a state that the highest animal species could think. They were conscious of their own past, present and future. They could organize parties, and invent luggage with wheels. They could also fight with hideous cruelty, and rig wholesale power markets. It took more than four and a half thousand million years for this to happen, but it was all very remarkable once it did. Some of the more advanced Thinkers took a look out into space, did a few sums, and worked out that there wasn't a cat's chance in hell of sentient life of such complexity on any other planet within light years of theirs. There was a chance, indeed, that there was nothing like the Thinkers anywhere else in all the known universe. Boy, did that make them feel special.

Some Thinkers made a spaceship that could blast perilously from the planet and land on a large rock in orbit around it. The occupants of the spaceship stood on the barren and lifeless surface of the rock and looked back at their planet. A small blue pearl in a thick sea of black mystery. That's how it looked. They were more than routinely glad to survive the return journey.

CARBON AND LIFE

The planet had not always been a blue pearl. It started out as a lifeless rock itself, with an atmosphere of unbreathable gas. But, in its particular solar system, it was just the right distance from the sun for something special to happen that no Thinker completely understands to this day. Somehow, somewhere on the planet, some atoms combined to form molecules that combined to form compounds that combined into strings that were able to replicate themselves. Maybe it was a great cosmic accident. Maybe the replicating chemicals dropped in ready-made on a meteorite, which is merely to pass the buck. Maybe one of the Gods that the Thinkers would later worship was responsible. Whatever, with this replicability of complex chemicals, life had begun. Hundreds of millions of years of our story had passed by this point. Life didn't, and doesn't, come easy.

The main building block of life on the planet was a chemical element called carbon. Together with water, it made up the single cells of the planet's early life forms. These cells eventually found a way to create energy that was so ingenious that many cells copied it without significant evolutionary improvements for several billion years. The cells took carbon dioxide, combined it with water, and built complex molecules called carbohydrates, giving off the gas oxygen in the process. All that was needed for this to happen was light – no problem with an enormous sun in the solar system – and a certain vaguely magical pigment in the cell. The Thinkers later called this process photosynthesis. Having photo-synthesized, the planet's early life forms could then burn the carbohydrate in their cells to produce useful energy, releasing carbon dioxide and water again. A simple loop that created energy along the way. Clever or what?[1]

The oxygen produced by photosynthesis proved to be a good thing for the planet. It was a gas that wasn't toxic to carbon life forms, and which allowed them to produce energy when taken into cells in different ways. Slowly oxygen built up in the atmosphere at the expense of carbon dioxide. The Blue Pearl was some planet. Its evolving life forms made their own breathable atmosphere.

Almost 4 billion years passed, and life evolved further, as it can, no matter what they might try to tell you if you go to school in Alabama. The single-celled early life forms found a way to form many-celled organisms with soft bodies. They didn't look great, and probably would have tasted even worse. Shortly thereafter, many-celled animals with hard bodies evolved. These guys looked even worse, like cockroaches and earwigs only much bigger. A kid could have given his sister a heart attack with these bugs.

Everything lived in the sea until this time, but eventually enough oxygen built up in the atmosphere for the first plants to appear on land. The first animals, by now equipped with primitive lungs to breathe the oxygen – so long as they spent most of their life submerged in mud – were not long behind.

Around 350 million years from the end of the story, forests grew thick upon the land of the Blue Pearl. They grew so thick, in fact, that dead trees and other plants built up seams of virtually solid carbon. These seams became buried by sediment, and hardened into a black shiny rock. The Thinkers would later call this phase of the planet's history the Carboniferous period, and they called the rock formed in such volume in that period coal.

APOCALYPSE ONE

Evolution was now rampant, and the diversity of animals and plants was huge. But then, a hundred million or so years later, disaster struck. Something messed up the Blue Pearl's fragile coat of breathable air and swimmable seas so badly that life on the planet came close to being wiped out.

The Thinkers would later call this disaster a mass extinction. Some specialists among them crawled all over the rocks formed at the time with hammers, notebooks and magnifying glasses, working out from the fossils in the rocks exactly which species had bit the dust, so to speak, and which had survived. They found that 90 percent of all species had disappeared. These Thinkers disagreed about the cause of the apocalypse. Many thought a vast volcanic eruption had made the atmosphere unbreathable, or suddenly changed the climate. Others thought they had found evidence that a giant meteorite had slammed into the planet, with much the same effect. Maybe both things happened at the same time. Whatever the cause, there had been earlier phases of extinction since the hard-bodied bugs first appeared, but nothing like Apocalypse One.[2]

As you can tell from the name, there were more apocalypses to come.

The Blue Pearl slowly healed itself. Over tens of millions of years, evolution built up a very different stock of animals and plants from the survivors of Apocalypse One. The planet warmed up again. A group of giant lizard-like animal species evolved, with tails, claws, impressive teeth and unimpressive brains. They came to dominate both land and sea, and would later capture the imaginations of countless youthful Thinkers, making them very glad they hadn't had to share the planet with the giant lizards. These beasts were definitely not sister-scaring material. One look at their teeth showed you that the last thing many of them did for a living was eat vegetables.

THE GREAT UNDERGROUND COOK-UPS

The Blue Pearl had been home to a number of weird natural phenomena by now, and, more than 4,450 million years into the story, and 150 million years before its end, another one happened. Giant blooms of microscopic plants formed in the planet's seas. These plants lived, photosynthesized, reproduced and died in many billions of billions. When they died they created a submarine rain of organic matter into the sediments on the sea floors. The rain was so thick that all the oxygen in the bottom waters was used up, and the organic matter was not oxidized as happened usually in the seas. In some places the sediments built up thick and fast, and their weight created enough pressure and heat to cook the organic matter, turning it into something called oil.

The oil was so light and fluid that it could flow upwards within the sediments whose weight had cooked it, even though that same weight had turned most of the sediments themselves into rocks. The oil was able to seep through cracks and pores in the rock. It mostly found its way back into the sea, or on to the land, where it turned into tar which bacteria slowly ate, leaving no trace. In some places, though, the oil couldn't get through the rock layers above and became trapped underground. This trapped oil became pressurized as more and more oil seeped up to join it from below. The Thinkers would refer to the places where oil was trapped underground as oilfields.

Another 60 million years passed, and evolution proceeded apace on the planet's continents and in its seas. Around 90 million years before the end of the story, there was a second major phase of oil creation, seepage and containment below the seas. The amount of carbon involved in these two great episodes of underground cookery and oil entrapment was measured in hundreds of billions of tonnes. Oil formed underground at various other times in this way, but never in anything remotely like these quantities. Gas also formed, both in the two phases of oil creation and at other times. It formed from the cooking of plants – microscopic and otherwise – in the sediments, and from deeply buried oil.

Another 25 million years passed, and the planet remained dominated by the giant lizard-like beasts, on land and in the oceans. They evolved ever more impressive horns, scales, teeth and tails. More and more species of animals and plants evolved alongside them. Believers in a God would have looked at the Blue Pearl at this time, in the full glory of its manifold biodiversity, and thought, Wow, but this guy knows a thing or two about design. If so, it's probably just as well He or She didn't give the giant lizards big brains, or else they might still be around today.

APOCALYPSE TWO

Just as before when biodiversity was setting new records, disaster again befell the Blue Pearl. I'm sure that's just a coincidence. Either that, or God has a strange way of clearing house. Around 65 million years before the end of the story, a giant meteorite slammed into the planet. It was only around 10 kilometres in diameter, but it hit with the force of tens of thousands of nuclear bombs.

The Thinkers don't call such things "weapons of mass destruction" for nothing. Once again the atmosphere clogged up. Once again, the climate changed from nurturing to unliveable. Around 50 percent of animal species bit the dust this time, including the mega-lizards.

Much smaller, generally hairier, animals survived the climatic disruption and took over the suddenly depopulated neighbourhoods. These were species who liked to steal the lizards' eggs and eat them, meanwhile keeping their own eggs within their bodies, or more exactly the bodies of the females of their species. Many millions of years passed again with relative tranquillity, and the hairy ones evolved into larger and larger species. Never again, though, was the planet home to animal species so large as in the era of the giant lizards.

THE ARRIVAL OF THE THINKERS

The hairies mostly had four legs, and continued to have for many more millions of years. But then, around 7 million years before the end of the story, two-legged creatures evolved. These animals were stooped and hairy but they could think in the sense that they could take a stick and use it to dig up insect grubs. Then, around 2 million years before the end of the story, more upright-standing, less hairy species evolved. Pretty soon these guys could go well beyond sticks and grubs. They quickly evolved grunts into speech and discovered that by rubbing two sticks together you could get fire. The Thinkers had finally arrived.[3]

The Thinkers appeared on the planet some 99.997 percent of the way into the story. If you were to think of the planet's history as a 24-hour clock, they would have appeared one second before midnight. But now, with the discovery of fire, they began to dominate the planet in a way no animal had come close to doing before. They began chopping down trees to burn for fuel and creating land to cultivate, a pattern that continued through to the end of the story. The Thinkers came to call their species *Homo sapiens*, and their planet Earth.

They were an aggressive animal from day one, and their thinking – impressive as it grew to be in so many ways – started out shaky where it came to their own security, and stayed that way until the end of the story. Unlike almost all other animal species, they would extend their aggression to the point of actually killing each other. The vast majority of animals on the planet would shake a horn or two at competing members of their own species, and that would be it, no matter how gorgeous the female they were usually in confrontation over. This was the case even if they were accomplished at ripping members of other species to shreds. But not the Thinkers. Conflict avoidance was most certainly not their strong suit. At first, they organized into tribes, and periodically killed the members of other tribes. Later they organized themselves into city states, built amazing cathedrals and opened fast-food joints that served food wrapped in leaves. But despite inventing athletics as a potential substitute for war, the Thinkers still couldn't think their way out of regular armed conflict, and their means of waging war grew steadily worse. City states under attack would do things like pour boiling oil on their attackers, and the attackers would do things like lob dead donkeys over the city walls with giant catapults, hoping to spread deadly diseases. Efforts to build their own weapons of mass destruction appeared early in the Thinkers' history. Later still, the city states organized themselves into nation states. The Thinkers drew and redrew lines on maps to show where their nation states began and ended. Mostly they redrew the lines by waging wars of increasing scale and bloodiness.

The Thinkers always struggled with the issue of security – they seemed unable to grasp the idea that they might best be able to build their own security and indeed survival by making sure their neighbours felt secure – but they had a host of good thoughts about other things. Some Thinkers began a long-running collective train of thought about science. Other Thinkers created sublime art. Pretty much all of them liked to have a good dance.

When they weren't waging war against each other, and sometimes while they were, the nation states started to trade. For hundreds of years, the Thinkers plied their trade in sailing ships made of wood. These ships ventured further and further afield in search of stuff that would turn a sovereign or two, eventually going right around the world, on which voyages half the ships' crews often succumbed to tropical disease and simple malnutrition. The undermanned ships struggled back to their home ports laden with such things as spice. Spice was hot, and sold for top sovereign.

Trade came at a price. The sea being a stormy and unpredictable place if you didn't have satellite photos of weather systems, ships all too often sank. This became a big problem, bottom-line-wise. One day, in a port called London, a group of Trading Thinkers worked out a way to make the loss of a ship easier on the balance sheet. They sat in a café on stools made of traded wood, drinking traded coffee from traded china cups, and worked out a way they could all share the risks of long-distance trade. The idea was that they would each pay a premium to create a pot of money. From this pot, traders who lost their ships would be reimbursed to the full value of the ship and cargo, though not of course the lives of the wretched sailors, who were two a penny. As the traders put it, "the losses of the few could be spread among the risks of the many". They called their idea insurance. It appeared on the scene three hundred years before the end of the story.[4]

THE THINKERS GET INDUSTRIAL

Not long thereafter, the Thinkers worked out that they could burn coal instead of wood, and soon after that they found a way to burn coal in steam engines that could power ships much better than sails could. The engines could also drive machines on land. Now the Thinkers could build factories, and generate some serious wealth without having to go off and find spice on the other side of the world. They had created what they called an industrial revolution.

The poorer Thinkers who staffed up the disease-ridden ships fared little better in the new industrial age. Many of them were dispatched underground to hack out the coal that had formed all those millions of years ago, in the planet's Carboniferous period. This was not a safe undertaking at all. Many, many Thinkers died as rock faces collapsed, water flooded into mines, or invisible and odourless methane gas built up and then exploded. If they survived their job, many would spend what remained of their short lives fighting for each breath, their lungs clogged with dust. This was allowed to happen for almost two hundred years, right to the end of the story. Like I said, the Thinkers were OK in some areas of thought, and not so hot in others.

Just over a hundred years before the end of the story came the amazing discovery that coal, as well as being used for steam engines and factory machines, could be burnt to generate steam that could turn a wheel to make something called electricity.[5] At the flick of a switch you could make it flow to power up just about any useful device, and quite a few less-than-useful ones too. Special coal-burning power plants were built to make electricity. Coal was burnt in them by the thousands, then the millions, then the billions of tonnes.

But coal was dirty. It wouldn't burn completely, and left soot everywhere, clogging up machines and making it very difficult to get your sheets dry on wash day. Just as this was in danger of becoming very irritating indeed, some Thinkers found they could drill below the surface of the planet, pump oil, and burn that.[6] They didn't have to send children down to get oil. It came bursting out of the ground under its own pressure much of the time, and made much less soot when you burnt it. The Thinkers set about burning oil *en masse* instead of coal.

They found it well suited to another new invention, called the horseless carriage.[7] This name arose because most Thinkers at the time got around by sitting on four-legged animals called horses, or in carriages pulled by horses. The horse-borne Thinkers laughed uproariously when they saw the first horseless carriages. But after a bit of value-engineering and some good old-fashioned marketing by the inventor, the horse men were soon standing in line waiting to buy one. This was a bad time for horses, because the Thinkers had never been too hot on pension rights for themselves, let alone their beasts of burden.

The horseless carriages proliferated like wildfire, just as the electric power plants had, and highways spread the length and breadth of the Nation States. Oil was soon being burnt in hundreds of millions and then billions of tonnes, just as coal had been.

The Thinkers then found they could also burn natural gas instead of coal, venting far fewer fumes. A pattern emerged whereby the nation states that were more advanced with their industrialization would burn gas for preference in power plants, and oil for preference in transport. The less advanced nation states would burn coal because they had to, and oil whenever they could get it.

By the time the story ended, the Thinkers were burning more than 8 billion tonnes of oil, gas and coal each year.[8] Many of them thought that everything was A1 OK. But, as we have seen, they had a tendency to think only selectively, and their Big Burn-Up was no exception. The Thinkers had overlooked two very, very big things.

BIG OVERSIGHT ONE

The problem was that carbon dioxide was inescapably created when oil, gas and coal burned, and carbon dioxide trapped heat in Earth's atmosphere. The carbon dioxide worked like a greenhouse in the atmosphere, letting light from the sun in, but trapping that same light when it was reflected back from the surface of the planet. Ever original, the scientific Thinkers called carbon dioxide a "greenhouse gas".

This characteristic of carbon dioxide had been known about for more than a hundred years, but on the principle of selective thinking nobody much worried about it. Eventually though, around sixteen years from the end of the story, a group of scientific Thinkers got together and decided to put that right, because they had been looking at their charts, and noticed that – oops, surprise, surprise – the average temperature at the surface of the planet was going up. Their observations applied to both the lower part of the atmosphere and the surface of the sea. They spoke out just at the time that a particularly dire drought hit the Number One Nation State, turning its agricultural breadbasket into a dustbowl and its prettiest national park into a giant bonfire. So quite a few Thinkers listened. This was another trait of the Thinkers. Once things became pretty obvious, they were capable of noticing them. In fact, within a short time, almost all the nation states on the planet – 167 of them – got together and decided to negotiate a treaty that would do something to cut back use of oil, gas and coal.[9] They really didn't want to risk burning their own home, as it were. Or flood it, because the scientific Thinkers had also noticed that the seas had started rising, and knew that if the planet warmed up they would rise much more, inundating all the coastal areas, where most of the richest cities and much of the agriculture and industry were located.

But there was a problem. More than a hundred years of uncon-strained burning of oil, and two hundred years of coal, had created quite a set of vested interests among the Thinkers. Many organizations were in danger of being left with what they called stranded assets, and this included not a few nation states, especially the ones where the oil had accumulated in large quantities during the two Great Underground Cook-Ups. These vested interests had created a web of power that transcended the throw-weight of the nation states. This web had become, in effect, a kind of empire. The Empire of Oil was loosely bound, and even capable of civil war, but it was without doubt the most powerful interest group on the planet. The biggest companies were all oil companies or companies making horseless carriages, in which most of the oil wound up being burnt. The single biggest user of oil, however, was the military of the Number One Nation State. They and the oil states dug in to defend their perceived interests, and elected not to hear the scientific Thinkers' warnings about the planet, or at any rate not to act on them.

Coal had also built up a powerhouse of vested interests. This was particularly difficult to understand, given the enduringly awful conditions of the workers who slaved to get the coal out of the ground.

So time went by. Leaders of the nation states gathered regularly to talk about the problem of the creeping greenhouse-gas build-up, but their negotiations produced much hot air and little action. The Empire of Oil made sure of that. They knew a thing or two about how to obfuscate, lie, and get away with it. In many ways, the coal lot stooped lower than the oil lot at this game, for some reason.[10]

The majority of scientific Thinkers saw this orgy of short-term self-interest unfold, and became alarmed. They had made some rather fancy calculations using their most powerful computers and found that the global thermostat would go through the roof if oil, gas and coal continued to be burnt in such vast quantities. They were becoming more certain of that as time went by. They worked out that the warming planet would cause trees to die in vast numbers, soils to dry up, and ocean circulation to slow down: that these and other amplifying impacts of the rising thermostat would add even more heat-trapping gases to the air, trapping even more heat, in a sort of snowballing fry-up that held the danger of not being clear enough to persuade the Empire of Oil to act until it was too late. What they were describing was a potential dooms-day machine, though they were much too polite to describe it as such. Some of the scientific Thinkers became shrill in their warnings. Guys, they said, let's not mess around here. We could be talking Apocalypse Three if we allow this to happen.[11]

The few scientific Thinkers who counted bug and plant species were even more polite than the scientific Thinkers who studied the climate, but they chimed in at this point. There were only a few of them because, well, you know ... bugs, plants, who likes to study that kind of stuff at school? Anyway, the bug and plant guys pointed out that, in fact, Apocalypse Three was already well under way, even without the imminent cooking of the planet. All the Thinkers had to do was keep on chopping down the forests and polluting the seas, and they would soon have wiped out as many species as the giant meteorites did.[12]

Really? asked the Thinkers with the temperature charts. Are things that bad? Then, er, forget the bugs and plants. View the problem another way. All this wealth we have built up with all these hundreds of years of industry and trade? We'll blow the whole lot away. We'll wipe out our coastal cities, and we'll destroy most of the wealth we have spent a century creating.

They weren't alone. Since the days of the coffee-shop traders and the wooden ships, the insurance business had grown into something even bigger than the oil and gas business, in terms of annual takings. Although the insurance industry had nowhere near the same balls as the oil industry, being run mostly by ultra-cautious Thinkers, some insurance types spoke out. They said they agreed with the scientists. Global warming would destroy their industry, they ventured, unless oil, gas and coal burning were cut dramatically. And without any ability to insure, how did the rest of the traders think they were going to make any more money? Or even maintain the value of what they had?

NEGLECTED ALTERNATIVES

Faced with increasingly irate scientists, many very worried nation states, and a few whistleblowing insurance-type Thinkers, the Empire of Oil finally wobbled a bit. Some Thinkers within it jumped ship and suggested that we really didn't have to burn so much oil, gas and coal after all. There were Alternative Ways to get energy that didn't produce heat-trapping gases. Most of these Alternatives used the heat of the sun, in one way or another, to make fuels and electricity.

The Thinkers had known that there were Alternatives for quite some time. But the Empire of Oil said more research work was needed before the Alternatives were ready for reliable use, and, meanwhile, wouldn't you like to have a tiger in your tank? Wouldn't you like to use Ultimate Fuel in your very own personal battle wagon? Wouldn't you like to use less choke with more poke?[13] So the vast majority of Thinkers were just somehow happy enough to keep right on using the Empire of Oil's products. In doing so they paid the Empire so much cash that it couldn't fail but become ever stronger.

The Thinkers had themselves a great big habit. The drug barons of the Empire of Oil rode the planet in executive jets, more powerful than any president except perhaps the president of the Number One Nation State. But then he was one of them anyway.

So despite the defectors from the Empire of Oil, and the growing dissent within it, little changed. The Great Addiction remained. The horseless-carriage makers rolled out a new class of product that drank a gallon of oil every mile when driven at speed.[14] The dimmer Thinkers, and there were plenty of them, queued up to buy it. The Thinkers had become a bunch of addicts hooked on a Class A+ drug that was going to screw them over big time, and the rest of the Blue Pearl with them.

That was even without the second Big Oversight.

BIG OVERSIGHT TWO

Big Oversight Two had to do with the Blue Pearl's two Great Underground Cook-Ups. The fact was that most oil had formed during two shortish periods in the history of the planet, and required very special conditions to be entrapped underground. Again, the Thinkers knew this, sort of. It meant that they knew that the oil to which they were allowing themselves to become so addicted would, well, kind of *run out* one day. Hardly anyone questioned this. They just didn't, you know, *talk* about it much.

Mostly this was because the Empire of Oil told them not to worry. Oil won't start to run out for a long time, its leaders said. There's plenty of time to keep burning, plenty of time to perfect research on the Alternatives and slowly bring them into energy markets. We're running into oil, not out it. There are forty years of supply at least.

Then, just a few years before the end of the story, some dissident Thinkers from within the Empire of Oil spat the dummy. They got together and said, Wait a minute here guys, play fair: oil is running out faster than this. These dissidents had done some sums, and reckoned that the days of easily accessible cheap oil were coming to an end some time soon. They calculated the point at which half of all the oil ever created and trapped would be used up, including not just the oil formed in the two Great Underground Cook-Ups but from any other times. They found that this "topping point" was just three short years after the end of the story, maybe even less. At that point, the dissidents said, Big Oversight Two would be clear to everyone. The Thinkers would no longer be able to run their lives and their industries on *growing* amounts of *cheap* oil. All they could expect thereafter were *shrinking* supplies of *expensive* oil. Gas would go the same way not long after, and wouldn't help with the yawning gap in their need for oil. They would look to the Alternatives to come to the rescue, but the Alternatives wouldn't be able to, or at least not in enough volume to make a big difference, because the

Empire of Oil and its Culture of Suppression had held them back all the years of the Great Addiction.

The Great Realization would then happen.

It would be a grim time, the dissident Thinkers warned. When it became clear that there was no escape from ever-shrinking supplies of increasingly expensive oil, there would be a paroxysm of panic. The Thinkers would face an energy crisis of mammoth proportions, and that, plus the panic, would spark an economic collapse of unparalleled awfulness. It would lead to hardship the like of which could only be created otherwise by, well, out-of-control global warming, actually.

Rubbish, shrieked the Empire of Oil. This one they *really* did not like. It was easy to see why. If it was true, or perceived to be true by the majority of Thinkers, a Great Reckoning would probably be just around the corner from the Great Realization. Because that was another collective trait of the Thinkers. When they got angry about something *en masse*, they *really* got angry. They liked the taste of vengeance. And their lawyers knew how to help them look back in anger.

And there ends The Story of the Blue Pearl.

Except, of course, it isn't really the end. It is merely the state of play on Planet Earth today. And in that sense, given the unfinished business, the ghastly stakes, and notwithstanding the four and a half billion years that it took to unfold, it is really only the beginning.

I
OIL DEPLETION

CHAPTER 1

Insufficiency

We have allowed oil to become vital to virtually everything we do. Ninety percent of all our transportation, whether by land, air or sea, is fuelled by oil. Ninety-five percent of all goods in shops involve the use of oil. Ninety-five percent of all our food products require oil use.[15] Just to farm a single cow and deliver it to market requires six barrels of oil, enough to drive a car from New York to Los Angeles.[16] The world consumes more than 80 million barrels of oil a day, 29 billion barrels a year, at the time of writing. This figure is rising fast, as it has done for decades. The almost universal expectation is that it will keep doing so for years to come. The US government assumes that global demand will grow to around 120 million barrels a day, 43 billion barrels a year, by 2025.[17] The International Energy Agency, the organization set up by industrialized countries to give them advice on oil and other energy matters, is scarcely less bullish. Its 2004 *Energy Outlook* forecasts 121 million barrels a day by 2030.[18] Few question the feasibility of this requirement, or the oil industry's ability to meet it. They should, because the oil industry won't come close to producing 120 million barrels a day. The most basic of the foundations of our assumptions of future economic wellbeing is rotten. Our society is in a state of collective denial that has no precedent in history, in terms of its scale and implications.

Of the current global demand, America consumes a quarter. Because domestic oil production has been falling steadily for thirty-five years, with demand rising equally steadily, America's relative share is set to grow, and with it her imports of oil. Of America's current daily consumption of 20 million barrels, 5 are imported from the Middle East, where almost two-thirds of the world's oil reserves lie in a region of especially intense and long-lived conflicts.[19] Every day, 15 million barrels pass in tankers through the narrow Straits of Hormuz, in the troubled waters between Saudi Arabia and Iran.[20] The US government could wipe out the need for all their 5 million barrels, and staunch the flow of much blood in the process, by requiring its domestic automobile industry to increase the fuel efficiency of autos and light trucks by a mere 2.7 miles per gallon.[21] But instead it allows General Motors and the rest to build ever more oil-profligate vehicles. Many sports utility vehicles (SUVs) average just 4 miles per gallon. The SUV market share in the US was 2 percent in 1975. By 2003 it was 24 percent. In consequence, average US vehicle fuel efficiency fell between 1987 and 2001, from 26.2 to 24.4 miles per gallon. This at a time when other countries were producing cars capable of up to 60 miles per gallon.[22]

Most US presidents since the Second World War have ordered military action of some sort in the Middle East. American leaders may prefer to dress their military entanglements east of Suez in the rhetoric of democracy building, but the long-running strategic theme is obvious. It was stated most clearly, paradoxically, by the most liberal of them. In 1980 Jimmy Carter declared access to the Persian Gulf a vital national interest to be protected "by any means necessary, including military force".[23] This the US has been doing ever since, clocking up a bill measured in the hundreds of billions of dollars, and counting.[24]

With such a strategy comes an increasingly disquieting descent into moral ambiguity, at least in the minds of something approaching half the country. The nation that gave the world such significant landmarks in the annals of democracy as the Marshall Plan is forced by its deepening oil dependency into a foreign-policy maze that involves arming some despotic regimes, bombing others, and scrabbling for reasons to make the whole construct hang together.

America is not alone in her addiction and her dilemmas. The motorways of Europe now extend from Clydeside to Calabria, Lisbon to Lithuania. Agricultural produce that could have been grown for local consumption rides needlessly along these arteries the length and breadth of the European Union. The Chinese attempt to emulate this model even as they enforce production downtime in factories because of diesel shortages and despair that their vast national acreage seems to play host to so little oil.[25]

This half-century of deepening oil dependency would be difficult to understand even if oil were known to be in endless supply. But what makes the depth of the current global addiction especially bewildering is that, for the entire time we have been sliding into the trap, we have known that oil is in fact in *limited* supply. At current rates of use, the global tank is going to run too low to fuel the growing demand sooner rather than later this century. This is not a controversial statement. It is just a question of when. One purpose of this book is to explain why.

Why, then, have we not been seeking an earlier transition to the alternatives that must lie beyond oil dependency? Hydrogen fuel, biofuels, fuel cells and advanced batteries are among the technologies that can provide the direct power for transportation in the future. Solar and many other forms of alternative energy can provide the electricity to split water into hydrogen and charge batteries. This too we have understood for decades. We have also known that there are massive untapped reservoirs of oil savings in energy-efficiency measures and innovative mass transit. These alternatives may not be able to replace oil quickly or easily, given their tiny current markets. But they work, and in most cases they have been waiting for a green light for years. This is without any further fruits of human ingenuity, suitably directed. In a society that put a man on the moon more than three decades ago, surely there can be no doubt that we could replace oil use if we seriously wanted to? I ask again, why have we not been fast-tracking the solutions to the problem long since? A second purpose of this book is to examine that question.

A third purpose of the book is to pose and endeavour to answer the question of how fast oil is now depleting. Finite resource that it is, there will come a day, inevitably, when we reach the highest amount of oil that can ever be pumped. Beyond that day, which we can think of as the topping point, or "peak oil" as it is often called, will lie a progressive overall decline in production. Putting the same question a different way, then, at the current prodigious global demand levels, where does oil's topping point lie?

This is a question, I contend, that will come to dominate the affairs of nations before the first decade of the new century is out, and one whose broad parameters I will outline now.

THE LATE TOPPERS VERSUS THE EARLY TOPPERS

A great battle is raging today, largely behind the scenes, about when we reach the topping point, and what will happen when we do. In one camp, those I shall call the "late toppers", are the people who tell us that 2 trillion barrels of oil or more remain to be exploited in oil reserves and reasonably expectable future discoveries. This camp includes almost all oil companies, governments and their agencies, most financial analysts, and most business journalists. As you might expect, given this line-up, the late toppers hold the ascendancy in the argument as things stand.

In the other camp are a group of dissident experts, who I shall call the "early toppers". They are mostly people who have worked in the heart of the oil industry, the majority of them geologists, many of them members of an umbrella organization called the Association for the Study of Peak Oil (ASPO). They are joined by a small but growing number of analysts and journalists. The early toppers reckon that 1 trillion barrels of oil, or less, are left.

In a society that has allowed its economies to become geared almost inextricably to growing supplies of cheap oil, the difference between 1 and 2 trillion barrels is seismic. It is roughly the difference between a full Lake Geneva and a half-full one, were that lake full of oil and not water.[26] If 2 trillion barrels of oil or more indeed remain, the topping point lies far away in the 2030s. The "growing" and "cheap" parts of the oil-supply equation are feasible until then, at least in principle, and we have enough time to bring in the alternatives to oil. If only 1 trillion barrels remain, however, the topping point will arrive some time soon, and certainly before this decade is out. The growing and cheap parts of the oil-supply equation become impossible, and there probably isn't even enough time to make a sustainable transition to alternatives.

Should the early toppers be right, recent history provides clear signposts to what would happen. The figure overleaf shows the history of the oil price since 1965. We will return to this history in some detail later in the book, but let me summarize its main themes now. There have been five price peaks since 1965, all of them followed by economic recessions of varying severity.[27]

Figure 1: History of the oil price – a rollercoaster of conflict and economic turmoil

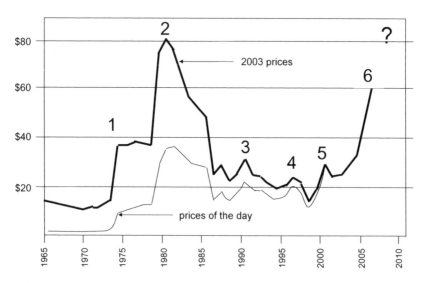

1. First oil shock, Yom Kippur War, 1973; **2.** Second oil shock, Iranian Revolution and Iran–Iraq War, 1979–1980; **3.** First Gulf War, 1990; **4.** Asian financial crisis, 1997; **5.** dot.com collapse, 2000; **6.** Third oil shock, 2004–?

The most intense peaks were the first two. The first oil shock, in 1973, saw the oil price more than double, reaching around $35 per barrel in modern value. The cause was an embargo by the Organization of the Petroleum Exporting Countries (OPEC), led by Saudi Arabia, and triggered through overt American support for Israel at the time of the Yom Kippur War. World oil supplies fell only 9 percent, and the crisis lasted only for a few months, but the effect was simple and memorable for those who lived through it: widespread panic.

The embargo was short-lived in large part because the Saudis feared that if they kept it up they would create a global depression that would cripple the Western economies, and hence their own. As it was, the short embargo created a miserable economic recession. I spent much of it doing my homework by candlelight. I didn't see much of my father. He was queuing for petrol.

The second and worst oil shock was triggered by the toppling of the Shah of Iran in 1979, and prolonged by the outbreak of the Iran–Iraq War in 1980. The first shock did not push prices as high as those at the time of writing, but the second shock pushed them to more than $80 a barrel in today's terms. Again panic reigned, even though the interruption to global supplies was only 4 percent.

The crisis ended when the price fell in 1981 for three main reasons. First, the Saudis opened their taps. With their huge reserves, mostly discovered in the 1940s and 1950s, they were able to act as a "swing producer", increasing the flow to bring prices down just as they had decreased it in 1973 to push prices up. Second, new oil came onstream from giant oilfields in more stable regions of the globe, including the North Sea. Third, large amounts of oil were released from government and corporate stockpiles.

These three reasons are high on the list of why we should worry today, because in the face of another shock things could not be resolved in a similar way. First, there are grounds to worry that the Saudis are pumping at or near their peak, no longer able to act as a swing producer, as we will see later. Second, the early toppers fear that there are no more giant oilfields left to find, much less wholly new oil provinces like the North Sea. Third, there is not much oil in storage, relative to current demand. The modern world works on the principle of just-in-time delivery. Our economies, overall, are more efficient in their use of oil than in the 1970s – a point much emphasized by late toppers – but the sheer weight of demand is much higher today, and it is still growing without an end in sight, or even strong governmental or corporate leadership demands that there should be one.

THE PEAK PANIC POINT

The cost of extracting a barrel of oil doesn't change much. A good rule of thumb might be $5 a barrel today, though obviously there are variations between oilfields in different geographic and political settings. What influences the price of oil most is confidence in supply and demand among oil traders. Oil prices are already at their second highest levels ever, in real terms, at the time of writing. Some pundits now profess that they will reach their highest ever levels, in modern value, within 2005.[28] This situation has arisen for many reasons, which we will look at later. But one of those reasons is manifestly *not* fear that the oil-production topping point is near. Early-topper arguments are not on the radar screens of the oil traders and analysts, as things stand. Should that happen, and should the mood of the packs on the trading floors flip to the view that we live no longer in a world of growing supplies of oil, but rather shrinking ones, the price will soar north of $100 a barrel very quickly.

An investor friend of mine has already concluded that this scenario is inevitable. He has switched his investment portfolio to anticipate the moment of "market realization". This peak panic point, as he calls it, will not be limited to oil traders. The worlds of economics and business routinely assume a future in which oil is in growing and cheap supply. Economists tend to assume that their "price mechanism" will apply. Higher prices will lead to more attractive conditions for exploration. This will lead to more oil being found, and the inevitable discoveries will bring the price down until the next cycle. Massive corporations write five-year plans based on assumed access to cheap oil and gas. Think, for example, how important such access must be to a chemical company dealing in plastics derived from oil. Or a food-processing company reliant on oil for every stage of food transportation, including of perishable final products, plus almost all the bottling and packaging and many of the preservatives and additives. But suppose the economists and corporate planners are wrong? Imagine the collapse of confidence when a critical mass of financial analysts, across the full breadth of sectors in a stock exchange, conclude that they *are* wrong?

If the topping point is indeed imminent, economic depression looms as a real prospect. The Saudis were right to be scared of this possibility in the 1970s. In the Great Depression of the 1930s, triggered in 1929 by the worst-ever stock-market crash, economic hardship was horrific. World trade fell by a breathtaking 62 percent between 1929 and 1932.[29] The widespread unemployment and social unrest bred Fascism in many countries, in some nations on a scale that would change the course of human history. As for the stock markets, it took them fifty years to regain their pre-collapse value in real terms.[30]

THE MULTI-TRILLION-DOLLAR QUESTION

To answer the core question of how close we are to the oil topping point, we need answers to a subset of three related questions. We can think of these as the "existing-reserves" question, the "reserves-additions" question, and the "speed-to-market" question. We will look at these questions in turn in Chapter 3. But, to study them with the best chance of full understanding, we need first to know something of how oil forms, is trapped, discovered and produced. So Chapter 2 is a deviation into the wonderful science of geology. For the reader, it will provide explanations for many of the events in The Story of the Blue Pearl. For me, it will be a trip back into my past.

CHAPTER 2

Finding oil

Geology is a science in which a lot of people can agree on basic truths with relative ease, both inside and outside the oil industry. But not all. I once spoke at length to the Ford Motor Company's lobbyist at the international climate negotiations. He tried, with conviction, to persuade me that the world was not four and a half billion years old but a mere ten thousand years.[31] He also told me that it didn't matter what we put into the atmosphere in terms of heat-trapping gases, because the Antichrist was coming, and along with his disciples – of which he wanted me to know I was one – the forces of God would be defeated in the Battle of Armageddon. And that wouldn't matter either, because the forces of God would then ascend to Heaven. He wanted to save me, get me to leave the Antichrist's payroll.

I'm sure he wasn't speaking for the board of the Ford Motor Company, but he scared me. The enormity and seeming unbridgeability of the gulf between his version of truth and mine was bad enough, but he assured me that he represented a massive movement in the United States, and that it was growing.

For my version of the truth, I prefer to go with the evidence of the fossil record and the tools of the earth historian. I studied for a PhD in geology at Oxford University and spent a decade thereafter researching earth history. Such a CV is by no means automatic qualification to be telling the truth, but it does mean I know the geological elements of The Story of the Blue Pearl to be true. My hope is that the reader will be in a position to judge how much of the rest of the story is true by the end of this book.

If you want not to worry too much about cooking the planet by burning oil and gas, or destroying economies because your employers are in denial about how much oil is left, then it obviously helps to think like the man from Ford. But I need to emphasize a very important point at the outset of this chapter. Most of the people in the business of pushing oil don't think that way. In the case of the geological history of the planet and life on it, as summarized in The Story of the Blue Pearl, my truth is generally the oil industry's truth. Almost all the people I know in the industry would accept the story of Earth's history as I briefly précised it as a parable in the prologue. The industry will fire many missiles at this book, hopefully all metaphorical, but not at the geological under-pinnings of it. How can they? That's a vital part of how they find oil. I know this, because I taught quite a few of them how to do it.

WHERE NOT TO FIND OIL: MOST OF THE PLANET

For most of the 1980s, I was a creature of Big Oil. I taught petroleum engineers and geologists at the grandiose-sounding but in fact quite tatty Royal School of Mines, part of Imperial College of Science and Technology in London. My researches on the history of the planet included such issues as the source of oil, and was funded by BP and Shell, among others. I also consulted for oil companies. In those days, I was psycho-logically insulated in a quest for the respect of my peer group, and highly selective as a consequence with the information I allowed on to my radar screen. The build-up of greenhouse gases in the atmosphere registered nowhere on my list of concerns. I had concerns about oil depletion, but only in the sense that cloaked my quest to find more with a certain nobility, at least in my own eyes.

The geologist has to find five things if he or she is to find oil. The first requirement is a "source rock", wherein organic matter can be cooked into oil. For this to happen the organic matter has to be buried to the depth at which the organic matter can be "cracked", as the geochemists call the cooking process. This depth must be no less than 7,500 feet and no more than 15,000 feet, as a rule of thumb. At 7,500 feet below ground the temperature is that of a cup of coffee, around 180°F (82°C). Given a good concentration of the right kind of organic matter at that depth for millions of years, oil will be the result. If the organic matter is buried to any more than 15,000 feet the temperature is more like that needed to roast a turkey (295°F or 154°C). In that kind of heat, the organic matter is cracked directly into natural gas.

Second there must be a "reservoir rock", into the pores of which the oil can migrate. The third requirement is a "cap rock", which stops the oil escaping from the reservoir rock. Fourth, there must be a "trap", a structure in the rocks – usually a giant fold – into which the oil migrates to be trapped under the cap rock. Finally, even if there is folding to create the trap, you need the structure to be leakproof. Too many fractures in the rock – features that geologists call faults – will cause the oil to seep out and not accumulate in drillable quantities.

This is the five-item tick-box checklist. It is an exacting one.

I know this.

10 May 1984, near Ormara, Makran Coast, Baluchistan
From the spine of the sandstone ridge, I can look far north across the rocky hills rolling into the haze towards Afghanistan. To the south, the Gulf of Arabia shimmers on the skyline across a dusty coastal plain studded with mud volcanoes, perfect scale models of Etna and Vesuvius tens of metres tall. I crossed this plain with my tribal guards this morning. The volcanoes have perfect model calderas at their peaks, the cool liquid mud in them bubbles as natural gas escapes. Smears of crude oil move in the mud like miniature slicks. Shaji, my colleague from the Hydrocarbon Development Institute of Pakistan, is excited. In the miniature slicks, he has seen with his own eyes that oil has been generated below this place.

Shaji gets excited about many things, it seems. The wide wadis that burst from the hills to cross the coastal plain have been in flood. Standing lakes of blue water now sit in them. Monsters live in the lakes, Shaji tells me. Half-eaten locals have been found by them. Rubbish, I say, and swim in the lakes at the end of each dusty day, while Shaji and the guards watch aghast, cringing and moaning to themselves.[32]

The Makran Coast fills with me with a feeling I have never experienced before: a mix of near-religious awe at the savage weird beauty of the landscape, and suppressed terror at its dangers. In this rocky desert Alexander the Great lost half his army to the elements while trying to get back to Babylon after his amazing ten-year voyage of conquest through the then known world. Flash floods I feel I can deal with. Snakes I know I can't. I walk around dressed like a Baluch tribesman in baggy salwar kameez and turban, but with knee-length Gore-tex gaiters above paratroopers' boots. I hope this get-up might be snakebite proof. I sit at night round the camp fire with my Baluch guards, looking like Lawrence of Arabia, just not as brave or well dressed. We talk via Shaji's translation. Their favourite topic seems to be how many Britishers their ancestors killed in the days of the empire, and how. Now their role is to protect me from traffickers who deliver Afghan heroin in convoys of Toyota Land Cruisers to boats on the coast. My guards carry single-shot pre-war Lee-Enfield rifles. The smugglers carry Kalashnikovs.

I soak in the view from the ridge. The planetary grandeur and the sense of insignificance it instils in me makes me want to shout aloud for some reason. But Shaji is thinking of oil. "There must be oil here," he announces, with a sweep of his arm. "It makes no sense that the Iranians and the Saudis have so much so close to this place."

Shaji is a worried man. His bosses in Islamabad are sure there is plenty of oil. An American oil company, Marathon Oil, came to explore here a few years earlier, drilled a few wells and left empty-handed. The aspiring oil bosses in Islamabad believed Marathon had found oil, but that the CIA had told the company to sit on the discovery. Everyone I talked to in Pakistan thought that.

As Shaji airs his frustrations, I inspect my measuring staff (it doubles as a cobra-probing pole) and say nothing. It is true that there are rocks all around us suited to the generation and entrapment of oil. But by now I am virtually certain there are no oil reservoirs worth drilling. At this point, I simply want to work out the geological history of this unique piece of the planet and earn the admiration of my peer group that way. I want to write an article in the Bulletin of the American Association of Petroleum Geologists.[33]

Before beginning fieldwork, I thought we might have all the five tick-box things you need for oilfields in the Makran Coast Range. For source rocks, we have mudstones rich in organic matter.[34] For reservoir rocks we have a thick sandstone formation. This rock looks good for the characteristics that are most important at the microscopic level if oil is to flow into a reservoir. It doesn't have a good name though: the Panjgur Formation, meaning the Five Graves. I never let four guards go out on fieldwork with me. For cap rock, we have more mudstones, impermeable enough in principle to hold oil in place in sandstones below.

We also have many of the structures needed if oil is to be trapped in the rock formations: big dome-shaped folds called anticlines. On satellite images of the Makran Coast we can see beautiful anticlines stretching for hundreds of miles along the coast.[35] Just around the corner, in the Zagros Range of Iran, north of the Straits of Hormuz, we know that similar anticlines are home to some huge oilfields. No wonder my Pakistani friends are frustrated.

What is missing lies in the structural geology. I knew there would be faults before I ever went to the Makran Coast, but not on the scale that we have found on the ground. There are simply far too many fractures in the rocks for oil to migrate along. It is vanishingly unlikely that any big traps formed intact. That, not the CIA, is why Marathon Oil quit.

I have discovered that, just around the corner from some of the biggest oilfields on the planet, an area the size of France almost certainly has not a drop of drillable oil.

And this is the case for most of the planet. Almost everywhere geologists have looked – which means everywhere by now, at least at some level of exploration – there is no oil because one or more of the five key requirements is missing. Even when all five boxes can be ticked, you can end up finding no oil. Only one well drilled in every ten finds oil. Only one in a hundred finds an important oilfield.[36] And as we shall see, the more wells that are drilled in a province or country, the smaller the oilfields generally tend to become.

WHERE TO FIND OIL: A FEW PRETTY UNIQUE PLACES, MOST OF THEM LONG SINCE LOCATED

Not far to the west of the Makran Coast, in the Zagros Range of Iran, and across the Gulf in Saudi Arabia, you have all the five prerequisites with golden ticks against each box. The source rocks are so rich in organic matter that all the giant oilfields of Saudi Arabia have been fed by a formation less than 100 feet thick. These rocks were formed in the Jurassic period: the time of the first Great Underground Cook-Up in The Story of the Blue Pearl.

Oil reservoir rocks are mostly either sandstones, made of sand grains, or limestones, made of calcium carbonate. In North America most of the reservoirs rocks are sandstone, and in the Middle East most are limestones. Much of the carbonate in the limestones derives directly or indirectly from shelly animals that lived in tropical seas; in many limestones, the fossilized shells of those animals are plain to see. Where there are no structures easily visible with the naked eye, much of the carbonate derived originally from sediment ingested by shelly animals and worms and ejected as faeces. The reservoir rock of the biggest oilfield in the world, the Ghawar field of Saudi Arabia, was made in this way: from a mass of what geologists might call faecal pellets but what others might call shit.[37] Ironic, isn't it?

A key thing to understand about oil and gas reservoir rocks is the way in which oil and gas seep into and through them. Without some appreciation of this, it is difficult to hold a considered opinion in the oil- and gas-depletion debate. I suspect that many economists, for example, have little appreciation of what follows. The two characteristics that geologists first look for in reservoir rocks can be thought of as "containment" and "delivery". Reservoir rocks are full of holes called pores. The amount of space in these pores is called the "porosity" and it defines how much oil or gas the reservoir rock can hold in principle – the containment. But a rock full of holes is no good unless there is good connection between them. Geologists call the size of the connections "permeability" – the deliverability.

The best cap rocks have zero porosity and zero permeability. The cap rock in Saudi Arabia is the best you could possibly find: a so-called evaporite formation. As the name implies, evaporites form in warm tropical seas when sea water evaporates. The first mineral to form in evaporating sea water is gypsum. The second is salt. All around the Mediterranean there are thick deposits of gypsum, formed when the Mediterranean waters evaporated. In some cases, despite water depths of 2 kilometres and more, the water evaporated entirely, difficult as that may be to conceptualize. In southern Sicily, in the spectacularly beautiful cliffs of Ericlea Minoa, you can see towering deposits made of giant clumps of gypsum crystals formed around 5 million years ago when the Mediterranean evaporated.[38] I once led a student field trip to Sicily and scrambled over these cliffs, mesmerized at the evidence of my own eyes. The rock was like a supersized crystal treasure trove two stories high. Above and below the gypsum layer were rocks that can only have formed in deep water. The Mediterranean really had dried up, and then flooded again. I stood there imagining how the waterfalls must have looked at the Straits of Gibralter as rising Atlantic sea water flowed back over to the natural dam in the straits to end the evaporative phase. It's funny how you remember small things in life. It was slightly late in the afternoon by the time we got there, and I recall a student looking at his watch and asking me in a bored voice if it was time to go for a swim yet.

The important point about gypsum for the oil industry is not its beauty or the miracle of its creation but that it is a mineral with a small amount of water in the crystal structure. Bury it to around a thousand feet, and that water is driven out, leaving another mineral, anhydrite. The porosity of anhydrite is zero. So too is the porosity of rock salt. Neither oil nor gas can get through it, so long as there are no faults. In the Arabian Peninsula, the anhydrite layer keeps the hydrocarbons locked tight in the fields, and the Swiss bank accounts brimming.[39]

EXPLORATION KIT: THE AGE OF NOTEBOOKS AND GEOLOGICAL HAMMERS IS LONG OVER

Geologists understood the significance of anticlines (those dome-shaped folds), when it came to the business of finding oil, more than a hundred years ago. Along the way, the search for them has expanded to more subtle types of traps that can bring source and reservoir rocks into contact with caps. Many of the people at the top of the oil industry today began their professional lives, notebook and geological hammer in hand, looking for anticlines with the right mix of rocks in far-flung areas of the world. By the time I explored in Baluchistan in the 1980s, I was roaming one of the last provinces on Earth scarcely to have been visited by geologists, but even there geological maps had been made before my time. The surface of the planet had been fully explored for oil by the 1960s.

Well logging: putting physics to work underground

Exploration since then has centred on two methods: subsurface geology and seismic exploration. When an oil well is drilled, geologists lower electronic tools down it and draw up a log of the rocks below. Such logs can be compared well to well, and in this way a subsurface three-dimensional map can be drawn up. The company that first developed the logging tools, Schlumberger, effectively found itself a licence to print money. The oil companies have outsourced their well logging to Schlumberger for decades.

The logs measure all kinds of physical properties in the rocks. What do oil and gas reservoirs look like in well logs? A big giveaway is to be found in resistance to electric currents. Water conducts electricity and oil and gas do not. Since 1920, heavy mud has been pumped into most wells to stop the oil rushing up the pipe causing a "blow out". Blow outs were a regular feature of the early days in the oil industry and are the devil's own job to bring under control, as any Texan roughneck will be proud to explain to you at length. The use of heavy mud, though, meant it was perfectly possible to drill through an oilfield and not realize you had done it. You needed your electrical logging kit to test for oil. Just after the war, a Shell geologist worked out how to use the logs to calculate how much oil and how much water there was in a reservoir rock. Measuring layer by layer through a reservoir, using all the exploration wells drilled into an oilfield, is one of the main ways that oil reserves are worked out.

Seismic exploration: shocking tactics

The second of the major exploration techniques involves bouncing sound waves into the rocks below ground and recording the reflected waves. This technique is called seismic reflection profiling. The sound waves are generated by airguns at sea and specially designed vibrating trucks on land. Strings of recording devices are trailed behind ships or laid out on land. Computers are then used to unscramble all the reflected waves from different depths and to create a two-dimensional picture of the subsurface structure. I remember watching the underwater booms of airguns trailed behind a ship when I worked off Japan in the early 1980s. Later, in the offices of the Japan Petroleum Exploration Company in Tokyo, I spent weeks trawling through profiles looking for indications of oil and gas. Hydrocarbons often appear on a seismic reflection profile as a "bright spot" in the subsurface picture. There were some of these offshore southern Japan, but they were too small to excite my Japanese colleagues or to be of much use to their energy-hungry nation.

The main point about seismic exploration is this. The oil companies first used the seismic technique in the 1930s, and by now every sedimentary basin has been explored to some extent, most of them with dense grids of seismic lines.

Today, the art has developed to such an extent that "three-dimensional seismic" is commonly used in the hunt. In this technique, closely spaced profile lines are projected by computers into a 3D subsurface image that can be manipulated on a computer by an oil explorationist in the comfort of his or her chair. It is easy to imagine how powerful the combination of 3D seismic and downhole logs can be in proving, one way or another, whether oil or gas might lurk below ground.[40]

EXTRACTING OIL

Let us distinguish at this point between conventional oil and unconventional oil. Conventional oil is liquid, and sits underground in a reservoir under pressure. In many occurrences, gas is trapped above the oil, exerting downward pressure, and water lies below the oil, exerting upward pressure. Unconventional oil consists of sands and shales containing solidified oil, or solid tar or bitumen deposits. Here the oil tends to look more like the black blocks of congealed goo cooked up by road repair teams than the slop that kills seabirds the day after a tanker hits the rocks. Like conventional oil, unconventional deposits are localized. This time the Middle East and Russia miss out: 87 percent of the deposits occur in Canada (36 percent), the United States (32 percent) and Venezuela (19 percent).[41]

Here is the sequence of events when conventional oil is found. First, a rotating drill bit on the end of a string of drill pipe, made of toothed cones, is aimed at a target underground. This can be a potential reservoir identified from seismic profiling, or a different part of an already proven reservoir. A pump pushes mud down the inside of the string to lubricate the drill bit, and to keep any oil and gas from escaping. In case that fails, the first few hundred feet of the well is lined with a thick surface casting held against the rock by cement. This casting contains valves called blow-out preventers that can be shut if oil and gas starts coming up the hole.

Once the hole is complete to the desired depth, the logging tools are lowered down to take their measurements. If they show indications of oil or gas, steel casing is run all the way to the bottom of the hole and held in place with cement. The founder of that well-known oil services company, Halliburton, made his first fortune selling these kinds of services to the oil companies. Shaped explosive charges blast holes one inch in diameter through the casing right where the oil is. The space in the well is then in contact with the pores in the reservoir rock, and the oil in those pores is free to move. In the best wells, the oil flows out of the reservoir and up the hole under its own pressure. It is driven through the holes in the casing either by the pressure of the water below the oil, or the gas above it, or by gas dissolved in the oil that separates out as bubbles when the field is produced, pushing the oil through the reservoir's pores.

Since around 1960, Halliburton and its contemporaries have been able to assist the process. They use truck-mounted pumps to jack up the pressure underground to the point where the reservoir rock fractures. This is called hydrofracturing. Some wells would not work without this technique.

When the field is ready to produce, pumps can be installed down the hole to lift the oil if it shows reluctance to move under its own pressure. The "nodding donkeys" you see in many oilfield shots work these pumps.[42] During production, water and gas can be pumped into the reservoir to push more oil out. This is called secondary recovery, but despite the name it can happen these days from day one.

The most oil these techniques can hope to push out of a reservoir in normal conditions is around 35 percent of the oil in place. Beyond that, "enhanced recovery" techniques can be applied. These include the injection of steam or detergents into the reservoir to mobilize the oil, the injection of carbon-dioxide gas or fizzy carbonated water to increase the volume of the oil, or an injection of air and the starting of a fire in the reservoir.

Another weapon in the oil industry's arsenal is called directional drilling. Believe it or not, holes can be drilled at all angles from one drilling platform. Bent holes can even be made to go horizontally. It is not unheard of for a well to be three miles long but to extend more than a mile beneath the platform. Using downhole sensors that show where his hole is going, the driller can effectively steer the bit with a joystick on a computer.[43]

Unconventional oil cannot be extracted anything like as easily as conventional oil. Canada's tar sands can be mined in open-cast pits where they are close to the surface. Giant shovels are used to excavate a mixture of bitumen, water and unconsolidated sand. This has to be carried to separation plants, where hot water is used to melt and separate the bitumen. Eighty percent of the deposits are too deep for open-cast mining, and extraction relies on processes such as steam-assisted gravity drainage, in which steam is injected below ground to melt the bitumen and coax it to the surface.[44] Whether open-cast or subsurface, the techniques involved are very energy- and water-intensive. We return to those problems in the next chapter.

CHAPTER 3

The topping point

We come now to our subset of three questions which need to be answered in order to establish how close we are to the oil topping point. These were as follows. First, the existing-reserves question: how much oil is there in discovered oilfields, mapped out, proved and ready to be exploited? Second, the reserves-addition question: how much oil remains to be added via new discoveries, enhanced recovery techniques and so-called unconventional oil? Finally, the speed-to-market question: how fast can the oil be delivered to fuel tanks, whether or not it is found to exist underground in reserves theoretically able to match the expected demand levels?

HOW MUCH OIL IS THERE IN EXISTING RESERVES?

Shell's big hint

In January 2004, the early toppers' case suddenly looked a good deal more worryingly feasible to those who have tended to take the late toppers at face value. Shell's then Chairman, Sir Philip Watts, told investors that the company had over-estimated its reserves by more than 20 percent. By March, internal e-mails had been requisitioned by lawyers and these made it clear that the Chairman and his Head of Exploration had known about this problem for some time, and had deliberately lied about it. Both men departed the scene. As I write, they and other directors face criminal prosecution in the United States.

Shell's corporate scandal is dramatic enough. But there is a clear risk that it is only the tip of an iceberg. Today, many people in the oil industry appear to be under pressure when it comes to supplies of oil. "There is something strange going on in this industry," Shell's replacement boss, CEO Jeroen van der Veer, told the press in November 2004. He suspects that other companies have the same problems he inherited. The *Economist* drew the following conclusion: "Industry analysts and investors are quietly saying that Mr van der Veer may be right, and another big reserves scandal may be brewing somewhere."[45]

BP and the "proved" energy bible

Against this unpromising start, how much oil do we think the oil companies have found to date? Call BP for a bit of help with the answer and you'll be sent their annual BP Statistical Review of World Energy.[46] In it, you'll see lists of data for national proven oil reserves. Add these up to a global total of oil reserves year by year, and you'll see the total creep reassuringly upwards over time. Figure 2 shows those figures, from successive annual reviews, plotted as a histogram, split into the Middle East and the rest of the world.

Global reserves rise from just over 600 billion barrels in 1970 to almost double that today: 1,147 billion barrels at the last count up to and including 2003.

So what's the problem? Hold off investing any of my pension in alternative energy until the 2020s, you might conclude. In fact, why not sign me up for an SUV while you're about it?

Figure 2: Global oil reserves; "BP's" view

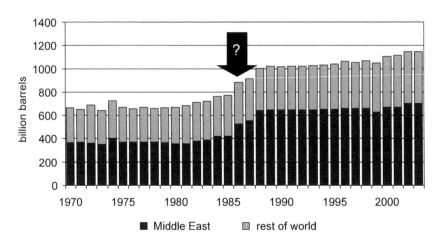

The first hint that something might be amiss comes, as is so often the case in life, in the small print. Squinting through a lens if you have anything but perfect eyesight, you will find that the data in BP's own report are not BP's at all. The estimates have been compiled using "a variety of primary official sources, third-party data from the OPEC Secretariat", and a few other places completely removed from BP's headquarters in St James's Square with all its accumulated research and knowledge. Think how many libraries of understanding BP must have gathered in over a century of aggressive oil exploration and production all over the world. And yet all they offer us as a guide to our own understanding of how much "proved" oil reserves there are left on the planet is a compilation of other people's data. And much of that itself is second hand.

After this revelation comes another. The small print continues: "The reserves figures shown do not necessarily meet the United States Securities and Exchange Commission definitions and guidelines for determining proved reserves, nor necessarily represent BP's view of proved reserves by country."

They don't even believe the figures they are publishing!

Referee! This is a publication used as an energy bible by researchers the world over! Students quote it as whole truth in undergraduate essays. Journalists quote it as gospel in legions of articles. They don't insert caveats like this! Neither have they seen such caveats in earlier reports.[47]

You might end up with a few questions for the authors of the BP Review at this point. But then, at the end of the document, we read the following: "BP regrets it is unable to deal with enquiries about the data in the Statistical Review of World Energy."

So what is BP's real view of "proved" reserves?

Could it go something like this?

Looking closer at the chart and zooming in, you'll see the figures show that global reserves of oil went up particularly quickly between 1985 and 1990. I've put a big black oily arrow to mark the point.

There must have been some big new oilfields discovered then, right?

Wrong. The actual new discoveries in that period were less than 10 billion barrels.[48] The Middle East nations hiked their "proved" reserves from *already discovered* oilfields by fully 300 billion barrels collectively in that period, professing one after another that their national calculations had all somehow hitherto been too conservative. Three hundred billion barrels is a lot of oil. It is more than a decade of demand at current levels.

Here's how it happened. In the 1950s, the nations with oil organized themselves into the cartel known as the Organization of the Petroleum Exporting Countries, or OPEC. OPEC's main aim was and is to try and control the price of oil. They don't want it too low. That would cut their income. Neither do they want it too high. That might get the addicts thinking of maybe going elsewhere. They want it just right, perhaps around $30 per barrel in today's money. To do this they can't produce too much, because that would flood the market causing the price to drop. They have to produce exactly the right amount collectively, and that means producing by quotas. After much bickering in the early days, the OPEC oil ministers decided in 1982 to allocate a quota to each country in the cartel according to the size of its national reserves.

But in 1985, they began to – how shall I put it? – massage the data. Kuwait, as it happens, was the first to give in to temptation. They found that their reserves had gone up overnight from 64 to 90 billion barrels. In 1988, Abu Dhabi, Dubai, Iran and Iraq all played the same card. Abu Dhabi had been so needlessly conservative that their reserves went up from 31 to 92 billion barrels! They surely must have employed some incompetent geologists. How could they have overlooked 60 billion barrels? Finally, in 1970, Saudi Arabia decided it too had been conservative. The desert kingdom hiked its total from 170 to 258 billion barrels.

You can also see in "BP's" data, in Figure 2, that the Middle East's reserves have been almost constant in size since then. What you don't see in the figure is that this is the case not just for the sum of the oil reserves of the Middle Eastern oil producers but also the figures of reserves for the individual nations.[49]

Consider the enormity of this coincidence.

It means that the billions of barrels found in new discoveries each year would have to exactly match the billions of barrels produced each year in each of the Middle Eastern OPEC nations, and do so consistently every year for more than a decade.

BP's Statistical Review of Everyone's World Energy Statistics Except Their Own invites us to believe all this without comment from them or recourse to questions by us. We are left to look at the total figure they cite for "proved" reserves, 1.1 trillion barrels, and think to ourselves ... "Er, really?"

Early-topper views

The early toppers have a different view. Being in most cases old hands from the oil industry, they know a thing or two about the games that go on in their industry. They estimate the total of proved reserves to be 780 billion barrels, some 300 billion barrels short of "BP's" figures. This is less than the world has produced since the first oil was struck over a century ago: 920 billion barrels by end 2003, a figure about which there is somewhat less controversy.[50]

Let us take some opinions that ought to be very difficult to discount, one from the very top of the oil tree in the US and two from the Middle East. Houston-based energy investment banker Matthew Simmons has been one of George W. Bush's energy advisers. He has studied reports by Saudi petroleum engineers showing that pressure is dropping in Saudi oilfields. The four biggest fields (Ghawar, Safaniyah, Hanifa, and Khafji) are all more than fifty years old, having produced almost all Saudi oil in the last half-century. These days, Simmons says, they have to be kept flowing in large measure by injection of water. This is of explosive significance, he argues. "We could be on the verge of seeing a collapse of 30 or 40 percent of their production in the imminent future. And imminent means sometime in the next three to five years – but it could even be tomorrow."[51]

The Saudis dismiss this, claiming that they have slightly more than the 258 billion barrels of "proved" reserves they claimed they had in 1970, with lots more yet to be found, and that they can lift the current extraction rate of around 9.5 million barrels a day to more than 10 with little difficulty. As Nansen Saleri, Manager of Reservoir Management at

Saudi Aramco, puts it: "… we have lots of oil, not only for our grand-children but for the grandchildren of our grandchildren".[52]

Saudi Aramco has the largest reserves of all the oil companies in the world: twenty times the size of ExxonMobil's, if they indeed have 260 billion barrels. They also have the lowest discovery and development costs, some 50 cents per barrel, or 10 percent of what the private companies pay in Russia or the Gulf of Mexico. And, being state-run, without much need for debt, they are under no particular pressure to divulge much to the financial markets. Lately, in the face of concerns about their ability to ramp up production, they have been marginally more open. They say they can maintain spare capacity of 1.5 to 2 million barrels per day and would be content with a fair price of $32–34 a barrel. Aramco's geologists have insisted they can hike output to 15 million barrels a day, adding more than 5 to the 9.5 reported today, 5 of which come from the giant Ghawar field alone. Contractors report that drilling activity is increasing, as indeed it needs to, given the age of the fields.[53]

But consider what A.M. Samsam Bakhtiari of the National Iranian Oil Company has told the *Oil & Gas Journal* about the existing-reserves question: "I know from experience how 'reserves' are estimated in major Middle Eastern and OPEC countries, and the methods used are usually far from scientific, as the basic knowledge for such a complex exercise is not to hand." Bakhtiari is withering about Saudi Arabia's reserves hike of 90 billion barrels in 1990. But he is not too keen on his own national figures either. The BP Statistical Review cited 92 billion barrels of "proved" oil reserves at the end of 1993, but Bakhtiari preferred the estimate of a retired NIOC expert, Dr Ali Muhammed Saidi, who could add the proved reserves up to only 37 billion barrels.[54]

Dr Mamdouh Salameh, a consultant on oil to the World Bank,

agrees there is a 300-billion-barrel exaggeration in OPEC reserves.[55] More recently, a former director of Aramco has said that Saudi Arabia's proved developed reserves stand at 130 billion barrels.[56] An anonymous informer talking to Dr Colin Campbell of the Association for the Study of Peak Oil goes further. His conclusion is that Saudi Arabia would go over its peak of production in the last quarter of 2004. This person speaks with front-line inside knowledge. "Saudi has at various times put nineteen fields into production," he says. "Of these eight are 'stars', being highly productive fields that produce around 90% of the country's production. All the others are 'dogs' that have never worked well and probably never will. Recovery rates of up to 50 percent may be appropriate for the 'stars'. For the 'dogs' 10, 15 or 20 percent would be more appropriate. Make this adjustment and Saudi has depleted more than 50 percent of its realistically recoverable reserves."[57]

In February 2005, Matthew Simmons speculated publicly that the Saudis may have damaged their giant oilfields by over-producing them in the past: a geological phenomenon known as "rate sensitivity". In oilfields where the oil is pumped too hard, the structure of the oil reservoir can be impaired. In bad cases, most of a field's oil can be left stranded below ground, essentially unextractable. "If Saudi Arabia have damaged their fields, accidentally or not," Simmons said, "then we may already have passed peak oil."[58]

HOW MUCH OIL CAN BE ADDED TO EXISTING RESERVES?

How much recoverable oil can we reasonably expect to add to global reserves in the future by new exploration, enhanced production, and extraction from unconventional oil? When we have an answer to that question, we can make an estimate of the original world endowment of oil: the total amount ever created and trapped on the planet over geological time. That is the figure from which the timing of the topping point can be calculated.

Let us start with a look at conventional oil in the United States. Where better to get a feeling about the pattern of oil depletion in the world than the country with the longest history of exploration, especially when that country is as large as America?

The pattern of historical oil discovery and production in the United States

Oil was first discovered in the United States in 1859. By the turn of the century, oil rigs could be found from Pennsylvania to California. Every potential oil-trapping anticline visible at the surface had been drilled by 1950. By then, Houston had emerged as the oil capital of Texas, America and the world. Shell set up a laboratory there, widely considered to be the best such in the business. One geologist who worked at Shell was a prickly man who insisted on being called M. King Hubbert. He may have been prickly because the M. stood for Marion. Marion Hubbert had become a legend in his own lifetime by 1956. One of his best-known discoveries was the mechanism whereby mountain-sized slabs of rock could slide for many miles atop thin layers of water-rich rocks needing only unbelievably light pressure from behind to do the pushing. This was one of the most amazing things I learned about as a geology undergraduate in the 1970s. One day in 1956, though, Hubbert had an idea that was not at all pleasing to his bosses. He calculated that oil production would peak in the "lower 48" states of America in 1971. He wanted to present a paper on his ideas to a meeting of the American Petroleum Institute. Right until the last minute his bosses at Shell were on the phone trying to pressure him into pulling it. But Hubbert was stubborn and not a man to be pushed around. He went right ahead and gave his paper.[59]

Hubbert knew better than most that oil forms only in select places where very particular conditions are met, and must be finite, both in the US and across the world. He knew that in other instances of finite resources, the pattern of exploitation tends crudely to follow the shape of a bell when plotted on a graph. He knew that individual oilfields follow a rising curve of production and go over a peak as the pressure drops within the field, followed by a descending arc of decline. We saw the reasons for this in Chapter 2. He figured that whole oil provinces, and therefore countries, would follow the same kind of bell shape. Early discoveries go slow for a short while, but as the resource enters increasingly common use and people learn how to find it the upside of the bell curve steepens quickly and rises fast. In the case of oil in the US, Hubbert knew the speed of that rise: the production rate had doubled every ten years from 1859 to 1956. Fifty-two billion barrels had been produced by 1956, and the nation was pumping at an unprecedented 2.5 billion barrels a year. But if the resource is indeed finite there must come a day when it is half used up. Critically, Hubbert knew that the annual rate of oil discovery in the United States had peaked in the 1930s. He couldn't see it going back up again once it started falling, and he figured production had to follow essentially the same shape.

To calculate when the peak of production would be reached, Hubbert needed an estimate of the total amount of oil that would ever be produced. Geologists call this all-important number the ultimate recoverable reserves. Twenty-five of Hubbert's eminent geological contemporaries had been polled on the issue just a few months before he gave his paper. Their estimates ranged from 145 to 200 billion barrels. Hubbert decided on 200 billion barrels. With this figure he could work out the volume under his curve. If each of his grid squares was, say, 10 billion barrels, his curve could only occupy twenty squares. He could come up with a pretty well-constrained estimate of where the peak was, assuming the estimate of the ultimate recoverable reserves was right. The date he came up with was 1971.[60]

What happened next?

Almost nobody believed Hubbert. Oil production was rising steadily at the time, and the whole thing seemed incredible. Never mind his standing as a world-class geologist based on his other work, many ridiculed the idea of Hubbert's Curve, as it became known. Shell censored the written version of Hubbert's address to the American Petroleum Institute, changing the wording of his conclusion to read that "the culmination should occur within the next few decades". The US Geological Survey, in particular, did everything it could to hike the estimates of ultimately recoverable American oil to a level that would make the problem go away. The US had 590 billion barrels of recoverable oil, the Survey said in 1961, meaning that the industry had thirty years of growth to look forward to.[61]

The years went by and the "lower 48" did indeed hit their topping point. It came a year ahead of estimate, in 1970, at 3.5 billion barrels. Since then, production has sunk down the second half of the curve at a steady rate. Many billions of dollars have been spent on ever more sophisticated exploration, including in areas where nobody imagined oil would be found at the peak of discovery in the 1930s, such as the deep water in the Gulf of Mexico. A frenzy of new domestic exploration began after the first Arab embargo in 1973 and the realization that domestic production could be ramped up no more. Every enhanced production technique invented has been tried and tested in American oilfields. But it has all made no difference to the remarkable symmetry of Hubbert's up-and-down curve. The US is just short of halfway down the second half of the curve now. In other words, it has used up some three-quarters of its original endowment of recoverable oil. Given its almost total lack of attention to the efficiency with which oil is burned, the US becomes more dependent on foreign oil imports by the day.

The US Secretary of the Interior at the time, Stewart Udall, later publicly apologized for having helped lull the American people into a "dangerous overconfidence" by accepting the advice of the US Geological Survey so unquestioningly. A long-serving US Geological Survey director who had led the campaign against Hubbert, V. E. McKelvey, was forced to resign in 1977.[62]

We need to remember this sequence of events, and the windows it gives us into individual and collective behaviour, when we come to consider the global oil topping point.

The pattern of historical oil discovery and production in the rest of the world

To what extent does Hubbert's Curve for American production apply to the rest of the world? The answer is that it is only a general guide to what is going on, and far from a rule. The US production pattern shows a quite smooth and symmetric curve because it was defined by very many actors operating in an unconstrained environment of free enterprise. As well as the giant oil companies, for example, there were always aggressive independent companies willing to take potentially high risks. In other words, if there was oil to be found in a potentially suitable area, there were always plenty of players with a chance to find it. Once found, the oil was pumped without much substantive effort at constraint. The curves for discovery and production are going to look different where conservative nationalized companies are doing the looking, or where – in the case of Saudi Arabia – there has been so much oil that the taps can be turned up and down for long periods so as to moderate supply and thus influence price. Countries that have onshore and offshore oil can have two curves, because the technology for offshore oil exploitation was developed much later than that for onshore. Curves will also be disrupted by big political events. War is not generally good for oil production in the countries where it is waged. Iranian and Iraqi production fell sharply at the time of their war in the 1980s. Collapse of regimes is another way to mess up a curve: the Russian production curve fell steeply when communism collapsed. Major accidents can even be expressed in the pattern production. The Piper Alpha offshore rig fire in 1988 severely disrupted the British production curve as companies scrambled to improve health and safety in the oilfields. Many countries show spikes on the upside of the curve, marking discovery early on of any particularly large oilfields, because they tend to be obvious prospects, drilled early. The issues discussed in Chapter 2 show why this might easily be the case. A giant anticline with a good source rock in the vicinity is not too hard to miss.

Despite these caveats, country after country follows a crude bell curve in both discovery and production. Today, more than sixty out of the sixty-five countries possessing oil have passed their discovery topping points and forty-nine of them have passed their production topping points. The US has a particularly long gap between the two: forty years (1930 to 1970). The UK has one of the shortest: twenty-five years (1974 to 1999). This is because the first discoveries were made much later in the UK than the US, when technology for both exploration and produc-tion were more advanced. Growing supplies of British oil didn't last long, though. Britain is now a net oil importer just like the US.[63]

Summing all the countries with oil, what does the global pattern of discovery and production of conventional oil look like? Perhaps unsurprisingly, it looks a bit like a crude bell, albeit a ragged one, as you can see in Figure 3 below. This picture tells us a very different story from that suggested in Figure 2, compiled from BP's Statistical Review of World Energy.

The historical figures for annual discoveries in Figure 3 were compiled by ExxonMobil, with reserves additions in known oilfields backdated to the year of discovery by the Association for the Study of Peak Oil.[64] The solid black line is global production to date. The dashed black line is the projected future extension of that curve according to the expectations of most oil companies and most governments. The figure plotted for "expectation", 43 billion barrels a year, is the US Energy Information Administration calculation of global demand in 2025, mentioned earlier.

I have numbered the vital parts of the story. Let us consider each of the numbers in turn, chronologically across the ragged discovery curve, and what those parts of the chart tell us.

Figure 3: Past global oil discovery and production

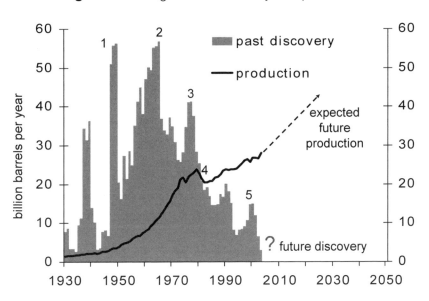

1. *The biggest oilfields in the world were discovered more than half a century ago, either side of the Second World War.*

    The big discoveries on the Arabian Peninsula opened with the discovery of the Greater Burgan field in Kuwait in 1938. At that time, it supposedly held 87 billion barrels. The slightly bigger Saudi Arabian Ghawar field, supposedly holding 87.5 billion barrels before extraction started, followed in 1948. These fields, the two biggest in the world, are so big that they dominate the global figures in their years of discovery.

2. *The peak of oil discovery was as long ago as 1965.*

    How many people appreciate this? I invite you to do a bit of personal market research. Line up ten of your better-educated friends. Preface your question to them with a few reminders about how many millions of dollars the oil companies make in daily profit, tell them an anecdote or two about the technical wizardry described in the last chapter, and ask them to imagine how many billions of dollars they must have spent on exploration over the years – both of their own money and the massive tax-deduction subsidies available to them. Then ask: in what year would you guess the most oil was ever discovered?

3. *There were a few more big discovery years in the 1970s, but none since then.*

    The biggest irregularity on the downside of the global discovery curve involved the discovery of oil in Alaska's giant Prudhoe Bay field, and the North Sea, in the late 1970s. I was a geology student then. I remember the thrill as the giant fields were discovered one after the other. They all had such serious-sounding names. Forties, Brent, Piper. I look back on those days now and I see something of the primeval attractions of the hunt in it. As a junior trainee hunter, I used to listen to the tales of the senior hunters, and how they had found their quarry, quite atremble with admiration. However, what I and the other hunters didn't know was that the days of giant discoveries were more or less over.

4. *The last year in which we discovered more oil than we consumed was a quarter of a century ago.*

   Since then, despite all those generations of eager brainwashed geology students, we have been burning progressively more, and finding progressively less. This is another one to try out on the ten educated friends.[65]

5. *Since then there has been an overall decline.*

   A small rise in discoveries in the 1990s that must have looked promising at the time has dropped in the opening years of the new century.

Does this sound like a world without a looming oil depletion problem, as BP CEO Lord Browne portrays it? Are people are being lulled into a sense of false security about oil supply based on his speeches, and publications like the BP Statistical Review of World Energy?

Reserves addition route 1: future new discoveries

Now we come to the first of the two bottom-line questions in the timing of the topping point. I have deliberately left the most important part of Figure 3 missing: an estimate for future discovery, as shown by the grey question mark. Looking at the best the oil industry could do for new discoveries in the opening years of the twenty-first century, where would you bet that the future discovery trend is heading? To achieve the level of production governments and most of industry expect in the years ahead, discovery clearly cannot continue on the same trend line it has shown since the 1960s. It must increase substantially.

Super-giants and giant oilfield discoveries: a pattern emerges
If you need to find a lot of oil in a hurry, it makes sense – as an oilman would put it – to look for elephants not mice. An elephant hunt makes sense of other reasons, particularly for the five "super-major" companies, ExxonMobil, Chevron Corporation (until very recently Chevron-Texaco), BP, Shell and TotalFinaElf. The more oil found at the end of the exploration process, the greater the return on capital invested. The super-majors cannot keep their reserves growing and their return on capital flowing fast enough to maintain a healthy share price if their explorationists find tiny oilfields. For this reason, a lot of hunting has been going on, and the results of that hunting provide the first reason for concern that we won't find a lot of new oil in a hurry, and can't. Let us consider the trend in discovery of "super-giant" oilfields, those holding over 10 billion barrels of oil each.

The Saudi Ghawar field, discovered in 1948, supposedly held 87.5 billion barrels before extraction started. The Kuwaiti Burgan field, discovered in 1938, held fractionally less. Given this start, how big do you think the third-biggest oilfield in the world might be, and when would you expect it to have been discovered?

Think of all that expertise that had been built up since the first oil was drilled in 1859. Think of all the trillions of dollars in oil revenues stacked up in the twentieth century, and all the hundreds of billions spent on exploration and the high-tech toys of exploration in the half-century since the biggest Saudi and Kuwait fields were discovered. Remember the sophistication of the seismic reflection profiling offshore. Consider the all-important oil source rocks, and how relatively limited they are in distribution. As BP's former Reserves Co-ordinator, Francis Harper, told the Energy Institute in November 2004, "We know how many world-class source rocks there are, and where they are."[66] Wouldn't it be reasonable to think that with modern technology at least one more field of more than 80 billion barrels might have been found somewhere, in all the places the companies have looked these last fifty years?

The third-biggest oilfield in the world is Samotlor, discovered in 1961, with 20 billion barrels. The fourth-biggest oilfield is Safaniya, discovered in 1951, at which time it also supposedly contained 20 billion barrels. The fifth-biggest oilfield is Lagunillas, discovered in 1926, containing 14 billion barrels. Only around fifty super-giant oilfields have ever been found, and the most recent, in 2000, was the first in twenty-five years: the problematically acidic 9–12 billion barrel Kashagan field in Kazakhstan.[67]

Let us reduce our scale of scrutiny from the super-giant to the merely giant. Half the world's oil lies in its hundred largest fields, and all of these hold 2 billion barrels or more, and almost all of them were discovered more than a quarter of a century ago.[68] Consider the recent record of discoveries of giant oil- and gasfields of over 500 million barrels of oil or oil equivalent. Half a billion barrels – the definition of a "giant" field – sounds a lot. But since the world is eating up more than 80 million barrels of oil a day at the moment, it is in fact less than a week's global supply. In 2000 there were sixteen discoveries of 500 million barrels of oil equivalent or bigger. In 2001 there were nine. In 2002 there were just two. In 2003 there were none.[69]

On the basis of this kind of evidence, is the industry going to plug the big grey question mark in Figure 3 with new discoveries? BP's

Francis Harper, for one, doesn't seem to think so. "Worldwide, the frequency of finding giant oil provinces and super-giant oilfields has been declining for decades and will not be reversed," he told an agog audience at a November 2004 London conference on oil depletion held in the Energy Institute. "We've looked around the world many times. I'd say there is no North Sea out there. There certainly isn't a Saudi Arabia."[70]

The very next day, by coincidence, I had a long pre-arranged breakfast meeting with someone much more senior in BP than Francis Harper, who had better go nameless. We had been colleagues of a sort in past years. Along the way in the conversation, I voiced my concerns about an early oil peak and asked for his reaction. "Don't worry," he assured me, "the oil peak is scaremongering. There is plenty of oil."

"You say this, and you should know," I replied. "But where?"

He didn't hesitate. "Saudi Arabia and Russia." No mention of deep-water oil or unconventional oil.

BP may be the biggest private oil company in the world, but it only produces 4 million barrels a day of the current 84 million barrels of global production.[71] Most of it is produced by inefficient giant companies owned by governments, Saudi Aramco being the biggest example. Assuming you can turn that knowledge into expanding BP production, I asked the BP executive, how can you be sure about the rest of your industry?

I already knew the answer. It was confirmed two weeks later when a very senior Shell executive visited my company to talk about solar energy, the meaning of life, and the oil topping point. He too had been a colleague of sorts, and is someone I admire greatly. He felt there was indeed a problem with conventional oil. "Big cats are becoming ever harder to find," he told me. For big cat read super-giant. But he too thought that oil supply could be made to grow. The solution for him was unconventional oil, especially in Canada, where Shell has operations in the tar sands. This prospect we will come back to in a minute, but first let us fill in the big grey question mark with the early-topper view of conventional oil production to come.

Figure 4: Past and future global oil discovery and production

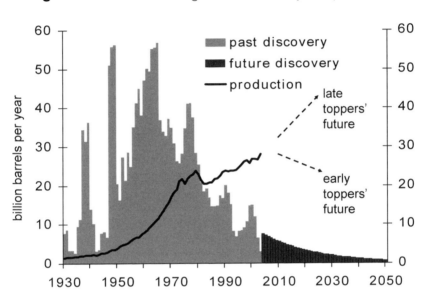

As we have seen, the early and late toppers differ by at least 300 billion barrels on how much "proved" recoverable oil awaits production underground. Little surprise then that they should differ by more than 1 trillion barrels on how much recoverable oil we can reasonably expect left to find. The ASPO projection, the downside tail extension shown in Figure 4, basically extrapolates the existing downward discovery rate to arrive at a total.[72] Knowing first-hand as they do of all the exploration history and technical capability described in Chapter 1, ASPO's dissidents argue that any super-giant fields would have been found by now, and therefore that any upward spikes and/or a long sideways skew in this tail-off curve are very unlikely. The total they come up with for oil we can reasonably expect to find is only 150 billion barrels.

Add this to the 920 billion barrels produced to date, the 780 billion in known fields, and ASPO arrives at a figure for global ultimate recoverable reserves of 1,850 billion barrels.[73] Some early toppers would go higher than this, with a common estimate of around 2,000 billion, or 2 trillion, barrels. Two trillion barrels would place the topping point between 2005 and 2010, depending on demand.[74] As Figure 4 shows, the topping point has a flat shape, so the exact day of the topping point is much less relevant than the day the oil traders and their clients realize that the late-toppers' future is unattainable.

Late toppers estimate up to 3.8 trillion barrels of remaining conventional oil. The US Geological Survey carried out a "world petroleum assessment" in 2000.[75] Undaunted by their track record on US domestic oil reserves in the years of opposition to Hubbert, the USGS conclusion for global ultimately recoverable reserves varied from 2,248 billion barrels to 3,896 billion, with a mean of 3,003 billion. Based on these figures, the US Energy Information Administration projects a peak in production as late as 2037, assuming growth in demand at an average of 2 percent per year.

Where would you bet your pension in this dispute, if you had to? As little as 1.8 to 2 trillion barrels or as many as 3 to 3.8? (Actually, come to think of it, many of you probably *are* having to bet your pension.)

To complete all the pieces of the jigsaw before we arrive at a firm view on the global production topping point, we need to consider deep-water oil, enhanced recovery, unconventional oil and gas. These I now consider in turn.

Deep-water oil: gold rush or hype?

In April 2001, a front cover of *Business Week* sported a picture of a tiger with a barrel in its mouth. "Exxon Unleashed", the headline trumpeted: "How the world's most powerful corporation plans to dominate the new era of oil exploration". Not to be outdone, that other battleship of the business magazines, *Forbes*, chose to focus on BP, showing the smiling face of Sir John Browne on its cover. "Black Gold Rush", shouted the title: "BP Amoco, The Hottest Prospect In The Oil Patch".[76]

A new era of oil exploration! A gold rush! The reason for all this excitement was deep-water oil. "We have opportunities today we could not have envisioned ten years ago," Exxon boss Lee Raymond crowed. Deep-water exploration held the key to rising oil output in the next decade, it seemed. Massive new fields lay in waters more than 1,000 feet deep, beyond the technology of a few years ago, but not that of 2001. Raymond likened the challenge to manned space travel.

The challenge was all the more important because BP had stolen a lead over Exxon and Shell in deep water. Sir John Browne had had much to do with this. When he was in charge of BP's subsidiary Sohio, he allocated the entire exploration budget to deep water, despite the fact that deep-water fields cost over a billion dollars to develop. "The key was to take a position in advance of the then-fashionable theory," Browne told *Forbes*. "It wasn't a 'bet-the-company' strategy, but it was clear that if it didn't work, our position in North America would be limited to Alaska." The gamble did work. BP started hitting sizeable deep-water fields in the late 1990s. By 2005, the company expected 25 percent of its production to come from waters 2,000 feet deep or more, up from 6 percent in 2001.[77] An old BP hand told *Forbes*: "Three times since I joined BP there has been a terrific buzz about the company: the North Sea, Prudhoe Bay, and now this. There is the sense that we are sitting on something huge."

All very optimistic and soothing, no doubt, to institutional investors' ears. But fast-forward to July 2004. A Merrill Lynch oil analyst, Ivan Sandrea, summarizes the state of play on deep-water oil in the *Oil & Gas Journal*. "Deep-water oil discovery rate may have peaked," he concludes. Since the mid-1970s, over 1,800 deep-water wells have been drilled in seventy areas worldwide. As of the end of 2002, 47 billion barrels of oil had been discovered. The peak of discovery was in 1996, at 5.8 billion barrels. Half of the discovered oil was in four areas: Brazil, the Gulf of Mexico, Angola and Nigeria, which analysts call the Big Four. The US Geological Survey, true to form, estimates ultimate reserves potential in excess of 100 billion barrels in deep water. Merrill Lynch's Sandrea thinks not. "There is no indication to suggest that three times the amount of oil discovered to date in the Big Four will be found again in these provinces, and outside the Big Four there is limited potential." Where does that leave us? "Global exploration potential looks more limited than ever," he concludes.[78] BP's Francis Harper agrees. "It is unlikely that another province the size of, say, deep-water Campos (Brazil) or the Niger delta will emerge," he says.[79]

ASPO estimates only 70 billion barrels of recoverable deep-water oil, and a peak of production of around 7.5 million barrels per day in 2014.[80] Sandrea estimates the peak at 6.2–6.4 million barrels per day in 2011–2013.[81] A big geological reason for the oil industry's disappointing results in deep water involves poor source rocks. Much of the sedimentary material deposited on the Atlantic continental slopes is lean in organic matter, and the organic matter is prone to be of the type that generates gas, not oil.

As we have seen, finding new oilfields – whether in frontier provinces onland or offshore – is only one of three ways additions can be made to reserves. The other two involve enhanced recovery and unconventional oil.

Reserves addition route 2: enhanced recovery in existing fields

The kind of improved oil recovery techniques we looked at in Chapter 2 have doubled the volume of production per well in the US since 1985. Reserves growth in existing fields has added more to remaining US oil than discovery of new fields has, despite all the hopes for deep offshore exploration. Horizontal drilling has had the biggest impact. Most of the 20,000 horizontal wells drilled worldwide as of summer 2003 were drilled in the US. Hydraulic fracturing and carbon-dioxide injection, as described in Chapter 2, are also in common use. USGS world energy project chief Thomas Ahlbrandt professes that potential global reserves growth in existing fields can contribute nearly as much new oil as discoveries of new fields yet to be made.[82]

The upper limit on recoverability of oil from reservoirs averages around 35 percent. The Saudis have reported that Ghawar's recovery potential has been lifted to fully 60 percent thanks to horizontal drilling and other techniques. Every 1 percent increase in global recovery factor would add about 55–70 billion barrels of reserves, almost equivalent to the North Sea. But BP's Francis Harper emphasizes a critical reason why the difficult and expensive process of enhanced recovery is unlikely to affect the timing of the topping point by much. "Around 200 billion barrels have been added to reserves between 1997 and 2003 from fields discovered between 1950 and 1996. But reserves' growth applies primarily to the bigger, older fields. The newer fields tend not to show reserves growth because we have become more efficient."[83] By this he means that, in more recent fields, enhanced recovery techniques have been designed in from day one.

A comprehensive survey of the sustainability of higher oil prices by investment bank Goldman Sachs in June 2004 took the following view. "The result [of investment in enhanced recovery] is that existing basins are declining much faster than in the past, requiring more capital just to keep production flat ... we estimate that the average age of the producing fields is thirty-six years and they are all concentrated around infrastructure that was built in the 'bubble era' of the 1970s."[84]

The Editor of *Petroleum Review*, Chris Skrebowski, puts the same argument more graphically: "Nobody sets out to develop an oilfield badly. The idea that there are lots of badly developed oilfields around the world awaiting American technology [to enhance their production] is a fantasy."[85]

Kenneth Deffeyes, an ex-Shell geologist and former colleague of Hubbert's, explains. He points to all the research that was done in 1980s, after the second oil shock, when billions of dollars went into enhanced recovery research, and much of the work proved successful by the 1990s. His conclusion? "That makes it difficult to ask today for new technology. Most of those wheels have already been invented."[86]

Such views are also found in OPEC countries. A.M. Samsam Bakhtiari of the National Iranian Oil company, mentioned earlier, reports primary recoverability in Iranian oil wells of 10–35 percent, and professes: "... seasoned Iranian experts seriously doubt anything near 50 percent could be achievable on most fields, even with all the technological paraphernalia."[87]

So what to make of the late-toppers' faith in reserves addition by enhanced recovery? Early toppers I have spoken with about this tend to view Thomas Ahlbrandt's assertion as the USGS repeating its track record of over-optimistic pandering to what governmental paymasters would like to think. Enhanced recovery has made precious little difference to the inexorable decline of US oil production, and it will be no different globally.

Worse than this, Matthew Simmons argues, enhanced recovery has actually created a "monster ball of depletion", that is accelerating the very problem others think it can help solve. Enhanced production in old fields today merely steepens the downward arc of the global production curve tomorrow.[88]

Reserves addition route 3: unconventional oil

Unconventional it may be, but there is an awful lot of it in place underground. The International Energy Agency thinks bitumen of various kinds may amount to 2.7 trillion barrels, up to 2.5 trillion of it in Canada, mostly in Alberta's tar sands. Alberta weighs in its tar-sand oil reserves at 315 billion barrels. The *Oil & Gas Journal* went along with that self-assessment in part in their 2002 annual summary of global reserves, promoting Canada overnight by 175 billion barrels, controversially, making it the second-largest oil-reserves holder. Venezuela has 1.2 trillion barrels of bitumen and heavy crude oil in its Orinoco oil belt, with an estimated 270 billion barrels classifiable as reserves. Oil-shale deposits total 2.6 trillion barrels, 160 billion of these classifiable as reserves. Adding all this up gives more than 7 trillion barrels of unconventional oil in place and more than 700 billion barrels of reserves.[89]

Seven hundred billion barrels of oil is almost as much as the early toppers think remains in conventional oil reserves. It is more than 60 percent of what BP's Statistical Review would like us to believe remains in "proved" conventional reserves. If that really is the case, all other considerations excluded, there would be much less to worry about.

But consider how much is being produced, the experience of production to date, what is planned as things stand in the years ahead, and what that tells us.

Tar sands: a lot of digging for not much paydirt

Canada's National Energy Board puts production from the tar sands at 1 million barrels a day in 2004. The Canadian Association of Petroleum Producers forecasts an increase in supply to 2.6 million barrels a day by 2015.[90] Not a lot of progress in ten years then. Even if all current expansion plans for Alberta's tar sands come to fruition they will produce only 3 million barrels a day by 2012. These are almost unnoticeable percentages of projected global demand.[91]

And even this has not been achieved without problems. The *Petroleum Review* reports "gargantuan cost over-runs" in the three projects operating at present in the tar-sands area, and "a crescendo of scale-backs and postponements" as a consequence.[92] This is because heavy oil is both difficult and expensive to extract. To yield a barrel of oil by mining tar sand which is exposed at the surface, two tons of sand have to be dug up, from which the oil must be separated in giant washing machines, and then huge volumes of tailings have to be dumped into giant sludge ponds. But much of the tar sand is underground. To get that out requires steam injection of high temperature (up to 200°C) and the use of solvents such as naptha, as we saw in Chapter 2. The environmental problems are only beginning at that point. Colossal amounts of water have to be heated, vast amounts of natural gas have to be used to do that, and it all involves major greenhouse-gas emissions.

The amount of water needed by the current small, unconventional oil-extraction industry in Canada is creating problems as things stand. Consider the following. Alberta's Environment Minister, Lorne Taylor, told a seminar on water management in June 2004 that the oil industry would eventually have to stop pumping water down wells for *conventional* enhanced oil recovery. Communities, farmers, ranchers and other landowners are becoming increasingly concerned about loss of fresh water, she said, and the province risked not having enough water in the future to sustain the population and protect the health of lakes and rivers. I emphasize, these concerns were about conventional oil production, not the much more water-intensive unconventional production techniques.[93]

Ironically, concerns are emerging that Canadian gas is depleting so fast that there won't be enough power available to heat water for the oil sands operations anyway. The *Petroleum Review* reported in November 2004 that the operations were using fully 0.6 billion cubic feet of gas a day. If production reaches 2.2 million barrels a day, the draw on gas could be as much as 2.5 billion cubic feet per day. The *Review* notes coyly that this would "place significant demands on dwindling supplies".[94] Canadian marketable natural-gas production in 2003 totalled 16.8 billion cubic feet per day, a decrease from 17.3 in 2002. Despite record gas-well drilling in 2003, production decreased by approximately 3 percent.[95] One critic told the *Oil & Gas Journal* that the amount of gas needed to extract the oil reserves that the magazine had so generously credited to Canada amounted to two or three times Canada's gas reserves. The only way the 300-plus billion barrels could be extracted, he figured, was to build nuclear power plants dedicated to the job. That might take rather too long for unconventional oil to fill the gap created by the conventionally depleting oil of the Middle East.[96]

The energy levels required to heat the water are such that carbon-dioxide emissions are more than three times greater than those of conventional oil production.[97] Canada has ratified the Kyoto Protocol, the United Nations treaty aiming to begin the process of cutting greenhouse-gas emissions. We will discuss this treaty and its seriousness further in Chapter 5.

To its credit, BP sees all this. "BP do not have any activities in Canada," Francis Harper says. "The main reason for this is the environmental damage and the main aspect of that is the CO_2 produced."[98] That laudable position contrasts strongly with Shell's. It will probably be difficult for BP to stay out of the unconventional pool for much longer, however, as the pressures to plug the gap build. Exxon has already begun planning a Canadian project with Imperial Oil.[99] CEO Lee Raymond told the October 2004 OPEC meeting that "with significant heavy oil, tar sands, and other 'unconventional' resources, new technology will be critical to making the 'unconventional' energy resources of today the 'conventional' resources of tomorrow. Making development of these unconventional resources economically attractive will ensure adequate supplies of fossil fuels are available at affordable prices for the next one hundred years."[100] Here, it would seem, is where the next new era of oil exploration lies, in Raymond's eyes.

The rest of the unconventionals: small beer

Of the 7 trillion barrels of unconventional oil in place in the world, Canada has the bitumen (36 percent of the total), and the US has the oil shale (32 percent of total). Most of that oil shale is in the states of Colorado, Utah and Wyoming.[101] In the case of oil shale, the optimism of the advocates is just as high but the problems of extraction are even worse. The term itself is a misnomer, because the organic matter in the shales is not organic carbon that has gone through the oil window, but kerogen, primarily, which must be heated to high temperature and processed in a complex and expensive process that throws up all the energy- and water-use problems of the tar sands and more besides. During and following the oil crisis of the 1970s major oil companies working on American shale deposits spent several billion dollars in various unsuccessful attempts to commercially extract shale oil.[102] As the next crisis looms, eager eyes are turning to them once again. In a March

2004 report, the US Department of Energy's Office of Naval Petroleum and Oil Shale Reserves actually concedes that an early conventional oil topping point is a real and present danger, and uses this an a reason for an emergency programme to keep the US, not to mention its navy, afloat and moving. The report concludes that 750 billion barrels of shale oil could be recovered ultimately. But it goes on to say that none of the requisite extraction technologies are proven at commercial scale, and even if they were, an aggressive production target for 2020 would be a mere 2 million barrels a day. An industry could not be initiated before 2011, because lead times for oil-shale projects from planning to commercial operation range from five to ten years.[103]

Where Venezuelan heavy oil is concerned, a key problem also involves the rate at which it can be produced. Unlike bitumen, it does flow, but slowly. Producing and transporting it requires special infrastructure, and lifting Venezuela's current production of 3.4 million barrels per day with heavy oil production will take too much time to have an impact upon the global oil production topping point.

BP's Francis Harper is dismissive. "Non-conventionals undoubtedly have a part to play in sustaining production but they cannot be the answer," he says. "I discount oil shale in my or even my children's children's lifetimes." The International Energy Agency currently projects all non-conventional oil growing at 8 percent per year to reach about 10 million barrels per day by 2030. "This is a Saudi Arabia, but a quarter of a century from now. Relative to conventional oil, this is pretty small beer."[104]

Can gas replace depleting oil?

Although there is strong potential for gas to displace oil-derived gasoline in the transport sector, we start from a point where a lot of gas is required just to keep the lights on and the water heated, never mind the vehicles running. Gas demand is expected to double by 2030, reaching 4.3 billion tonnes of oil equivalent per year. Over 40 percent of this projected demand will derive from power generation, a minority market for oil.

Gasfields deplete very differently from oilfields, gas being much more mobile than oil. It is normal for a gasfield to yield 70–80 percent of its gas over its production lifetime, whereas an oilfield will typically yield only 35–40 percent of its oil. Drillers normally set gas production far below the natural production capacity so as to give a long production plateau, but the danger in this is that the end of the production plateau comes abruptly, and without market signals.

As with oil, there are conventional and unconventional gas deposits. The conventional deposits sit in reservoirs as described in Chapter 2. Prominent early topper Colin Campbell estimates that the original global endowment of conventional gas was around 10,000 trillion cubic feet (equivalent to 1.8 trillion barrels of oil), of which about a quarter has been produced to date. He expects a global plateau in production of around 130 trillion cubic feet per year during the period 2015 to 2040, with production falling over a cliff beyond that. Jean Laherrère forecasts 12,000 trillion cubic feet for all gas including unconventional sources (2 trillion barrels of oil equivalent). He puts the peak of gas depletion in 2030, at 130 trillion cubic feet per year.

In the US, the situation is already not dissimilar to that of oil. As one American geologist told the *Oil & Gas Journal*: "… the gas production decline in the US is due to lack of good, large prospects. We have drilled more than 1 million gas wells in the US, and we are running out of significant targets." The flagship industry journal feels able to draw its own conclusion: "If there is an emerging consensus on the gas supply question, it is this: the US faces a near-term shortfall in gas supply, and there is little near-term solution in sight on the supply side." In the face of this, the US Geological Survey is less bullish on gas reserves than it is on oil. It foresees a total endowment of 15,000 trillion cubic feet (2.56 trillion barrels of oil equivalent), only 2,000 more than Laherrère. The US has used up 40 percent of its natural gas endowment and has less than 10 percent of remaining global reserves. There is lots of potential abroad, but 71 percent of global reserves are in the Middle East and the former Soviet Union. Russia and Iran alone have 45 percent. Gas has all the same problems of dependence on overseas supplies and more besides.[105]

Eighty percent of US electricity-producing capacity coming online is gas-powered, and US natural-gas demand will grow by more than 50 percent by 2025.[106] Gas imports from Canada by pipeline and liquid natural gas (LNG) from overseas are unlikely to meet this demand, much less replace oil in the energy mix. Four LNG terminals are up and running in the US, and thirty more have been proposed. But, as a recent analysis in *World Energy Review* concluded, "… importing more LNG will present more obstacles that all but a minority of industry optimists fear may be difficult to surmount. Chief among the problems are safety, NIMBYism, and terrorism." Historic explosions at LNG plants have killed more than a hundred people. Voters do not want regasifaction plants anywhere near them. Fire officials worry that any major fire set off on a thousand-feet-long LNG tanker by a terrorist suicide attack could destroy a city.[107] As Bob Ineson of Cambridge Energy Research Associates told the *Oil & Gas Journal*: "… demand destruction is the only real short-term alternative".[108]

At this point, some late toppers point to unconventional gas deposits, just like they do in the case of oil. And, as with oil, unconventional deposits are vast. They take the form of methane hydrates. Hydrates are solids made of water resembling ice that build up in sediments and trap natural gas as it is generated naturally by the breakdown of organic matter in the sediments. They are not ice as such because they can form above freezing point if the pressure is high enough. Most natural gas created in sediments simply leaks out of the sea floor into the ocean. Where the conditions are right for hydrates to form, gas builds up both within the ice-like hydrate layer, and – because the hydrate is impermeable – also as free gas below it.

The US Energy Information Administration puts the global amount of hydrate at 742,000 trillion cubic feet. But the vast majority of this is inaccessible to conventional drilling in very deep water. Where it is accessible, the deposits are very unstable. In my days as a geologist, I drilled off the coast of Mexico on one of the first scientific cruises to drill into a methane hydrate. We did it by accident. We were not allowed to drill into these layers in case we went through them or disrupted them in some way that would release the free gas below. Ships don't float too well on a mix of gas and water, and there was deemed to be real risk of ours sinking.[109]

The US Department of Energy is funding experimental drilling in hydrate regions, and the Russians and Japanese also have research programmes. However, as the *Oil & Gas Journal* concludes, "… no one expects exotic gas resources to make a significant contribution to global gas supply for decades".[110]

GETTING OIL TO MARKET

Here we reach the third and final of the questions we needed to address to answer core question about how close we are to the oil topping point: how fast can the oil be delivered to the fuel tanks and on to market?

The oil industry had to go further and further afield to find oil during the twentieth century. Giant oilfields were found almost from the outset. The first came in Peru, Texas, and Mexico. Then came the Middle East in the 1930s and 1940s. Then came Alaska and the North Sea in the 1960s, the latter not far from the markets but posing a whole new generation of challenges, with oilfields not only underground but under an angry sea. These days, very little oil is found close to the places where it gets burnt. Most is in fact found on the other side of the world from its major markets. The same goes for gas.

The problems were bad enough in 1967, when the largest oilfield ever found in North America was discovered on Alaska's North Slope. BP and the other companies involved in the 12.5 billion barrel Prudhoe Bay field knew that, short of shipping the oil round Alaska in specially designed ice-breaking tankers, a pipeline at least 800 miles long would be needed. It took eight years to build that pipeline, including delays for all the hearings needed to assess the environmental damage it would inevitably cause. The first oil from Prudhoe Bay didn't arrive at the shipping terminal in Valdez until 1977. By 1980, a full quota of 2 million barrels a day were flowing through the pipeline, but it had taken a decade for this to happen.[111]

From Valdez, the crude oil had to be shipped by tanker. The holds of oil tankers were by this time the size of cathedrals. When the *Exxon Valdez* hit the rocks in Prince William Sound in March 1989, almost a quarter of a million barrels of oil went into the water.

The modern equivalent of the Alaska pipeline is currently under construction between Baku on the Caspian Sea and Ceyhan on the Mediterranean coast of Turkey. A consortium of eleven Western companies, led by BP, are laying the 1,090-mile, 42-inch-diameter pipeline through earthquake-prone mountains with hostility-prone inhabitants: some of the most inhospitable territory ever encountered by the oil industry. Once the oil hits Ceyhan, it will then be transferred to tankers *en route* to oil refineries around the world. These final links in the chain that gets oil from a rock to a gasoline tank are familiar to seabirds and humans alike.

The point is this, in summary. It takes a long time to find new oil, a long time to produce it, and a long time to transport it to market. The production topping point is going to be defined primarily by oilfields already found, by how much is left in them, and by how much production can be enhanced.

CONCLUSION: THE TOPPING POINTS FOR ALL TYPES OF OIL AND GAS; OR A NEW TAKE ON "HOUSTON, WE HAVE A PROBLEM"

Colin Campbell has summarized the work done by the early toppers' umbrella group ASPO on the depletion of the different types of oil and gas reserves. He uses a "stacked peak" diagram, reproduced in Figure 5 opposite, to portray the result.

As we have seen, the definition of "reserves" is very much in the eye of the beholder. For Campbell and the other early toppers, "proved reserves" merely means "proved so far, and only if you can believe the original reports". Geologists commonly use the term "proved and probable reserves" for best estimates of what is most likely ever to be recoverable from an oilfield or gasfield. The figures plotted are ASPO's best estimate of proved and probable, plus ASPO's estimates of future additions to reserves from oil and gas yet to be found.

"Regular oil", as Campbell uses the term, excludes conventional oil from deep water (beyond 500 metres depth), conventional oil from polar regions and natural-gas liquids (liquid hydrocarbons derived from gasfields and produced by gas plants). It also excludes unconventional oil (also known as heavy oil and referred to as such in the caption). These are all plotted separately.

Figure 5: ASPO depletion curves for all oil and gas

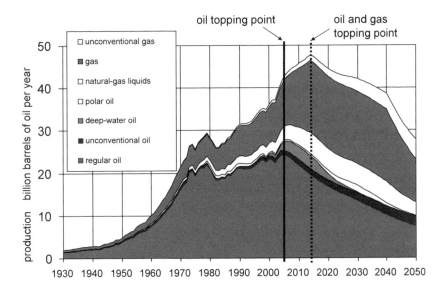

Here is the result.

Regular oil hits a flat topping point around 2005.[112] Adding oil from unconventional sources, deep water, the polar regions and natural-gas liquids defers the peak until around the end of the decade. Adding gas into the equation pushes the main peak for all oil and gas out to 2015, beyond which lies a steep descent on the other side of the curve.

Some early toppers favour a date for the oil topping point earlier within the 2005–2010 window, others a later date. Some can't or don't want to opt for a particular year, given the uncertainties and the flatness of the peak. Kenneth Deffeyes believes the peak is already upon us, but that we just can't see it yet. He opts for 2005 as the latest possible date.[113] Matthew Simmons won't be drawn on a year, but says peak oil is "likely at hand".[114] Chris Skrebowski favours 2008 plus or minus two years.

But in one sense the actual day on which the world manages to pump the most oil it ever can is academic. The key questions are these: Where will the peak panic point be in society, and in the markets on which our economies depend? And what happens when that day dawns?

CHAPTER 4

How serious is the crisis?

Occasionally, in the writings of the people who are pushing us into this mess, a little lack of self-confidence shows through. The US Department of Energy routinely publishes bullish accounts of how much oil we need and how much there is out there to burn. Its satellites, such as the US Geological Survey and the US Energy Information Administration, motor along in full horn-blowing support. But earlier I mentioned a report by a strangely named unit within the Department of Energy entitled the Office of Naval Petroleum and Oil Shale Reserves. This office seeks support for a major campaign to dig up and process America's oil shales, and points to the potential for an early oil topping point as a reason to do this. "A serious supply–demand discontinuity", the report contends, "could lead to worldwide economic chaos."[115]

There you have it. Later I will elaborate, but that – in a nutshell – is how serious it is.

THE CREDIBILITY OF THE KEY WITNESSES

Is there any chance that the early topping point of oil production is somehow wrong, all just a bad dream? I'm sorry to say I think not. It is important to realize that the early toppers are not advocates or agitators by choice. They tend to have high residual affection for the industry they have spent their lives in. They are, to some degree, whistleblowers.

Colin Campbell

The founder of the Association for the Study of Peak Oil (ASPO), Colin Campbell worked for forty years in the oil industry before retiring to western Ireland. He started his geological life with a PhD from Oxford University, as did I. We shared the same supervisor, a quarter of a century apart.[116] Campbell worked as an explorationist first for Texaco, then BP, then Amoco. He was Amoco's Regional Geologist for Latin America, then Chief Geologist for Ecuador, and later Exploration Manager for Norway, working in the North Sea. He then moved to Fina in Norway as Executive Vice President. I recite this CV to show a man who knows many of the oilfields of the world, and the community of people who have worked in the others. Now seventy-three, Campbell has the look of a High-Court judge, albeit somewhat in need of a haircut, and speaks in the measured cadence of a veteran BBC foreign correspondent. He likes to sit in his local pub explaining his depletion research to visitors like me with the aid of a half-full glass of Guinness. Essentially ignored in his whistleblowing when he first came out of his industry closet in the mid-1990s, in the aftermath of the Shell reserves fiasco Campbell has become a man much sought after.

A little context is needed before looking at Campbell's work and opinions. Accurate data on oil and gas reserves have been viewed as corporate and state secrets for a long time. US Securities and Exchange Commission rules for disclosure date back to the 1970s, and are widely criticized as vague and out of date. Different companies sharing the same fields can quote different reserves figures. The SEC, reviewing the situation in the wake of the Shell reserves downgrade, is encouraging companies to elaborate their accounting in the section of their accounts known as Management's Discussion and Analysis, or MD&A. However, it seems to be leaving this as a voluntary requirement, not a statutory one.[117] So secrecy spans most of the global reserves base, to varying degrees, as well as import and export statistics. An article in the *Financial Times* in December 2004 summarizes the situation as follows: "… in few industries do statisticians have to behave like spies, but those charged with collecting data on oil supplies have to rely on secret networks of informers". Despite growing calls for reform, "… the spies are likely to be in business for some time to come".[118]

At the end of his career, Campbell acted as consultant to an organization that co-ordinated an information network of this kind on oil, Geneva-based Petroconsultants. This is where his work on depletion took off. He worked on a database of information about the size and other characteristics of some 18,000 oilfields. It took men of his experience, and their networks of contacts, to do this, and Petroconsultants had been doing it for forty years. The only other information available at the time on reserves and production came from surveys carried out annually by the *Oil & Gas Journal* and *World Oil*. These trade magazines simply send questionnaires to companies and governments, and publish the unverified information that comes back. The oil companies of course wanted sight of Petroconsultants' database, and paid for the privilege. It was, and is, a unique resource.

In 1995 Campbell worked on the database with a French geologist expert in statistics, Jean Laherrère, formerly of Total. Campbell presented a report on their joint work to a conference in Cyprus. He argued that there was less oil than everyone assumed in supposedly proved reserves, and a lot less in ultimately recoverable reserves. Texaco sent someone to this conference. Their exploration manager immediately called Petroconsultants and told them that if the company wanted Texaco's business they should stop saying what Campbell had said. Thus began a new era in Campbell's life. He didn't stop speaking. Instead, in 1998, he and Laherrère published their work in the influential popular science journal *Scientific American*.[119]

The founder of Petroconsultants, Harry Wassall, ran his company like a "club for a network of impressive oil industry fugitives", as Campbell puts it. Campbell speaks with great affection of those days, and the relaxed atmosphere in which the ageing oilmen went about their work of compiling the data that many governments and oil companies would rather remained uncompiled. When Wassall died, in 1995, the company was sold to IHS, a database company. The new owners were more willing to say what the main client wanted to hear than the impressive fugitives had been. Harry's club was pensioned off, and the database was subsumed into IHS, where you can buy a licence to view it today for around a million dollars a year. Before you make the investment, remember that very few big oilfields have been found since 1995: 80 percent of world oil production today comes from oilfields discovered before 1973.[120]

Campbell has updated his own version of the database since 1995. In 2001, he founded ASPO, a loosely affiliated group of professionals from the oil industry and academia concerned about oil and gas depletion. The organization is headquartered in the University of Uppsala, holds annual conferences and publishes a monthly newsletter. Campbell's approach in adding to the database annually has been to take the average of the five public databases, which these days come from the *Oil & Gas Journal*, *World Oil*, the BP Annual Statistical Review, OPEC and ENI. All give different estimates. He then applies a factor based on confidential data sets to which he is given privileged access by key players in his old network. He does not reproduce their data. He is careful to revise any upgraded estimations based on enhanced recovery back to the date the original oilfield was found, so giving a true depletion picture. Campbell admits that this method cannot be as accurate as he or anybody else would like, given the seriousness of the issue. But, short of a remarkable rethink by governments and companies on transparency, an understanding of oil reserves and resources in the future will be forced to focus on private databases of this kind, where the key asset is the knowledge, experience and contacts of the group maintaining and updating the data.

What then does Campbell have to say about BP's Statistical Review, the most frequently used source of information on reserves? If we are running into oil instead of out of it as the Annual Review suggests, he asks, why are they and the other companies not investing in tanker capacity in line with projections of rising demand? Why, in the face of projected demand, are they allowing a refinery shortage to accrue? [121]

Chris Skrebowski

You might think the Editor of *Petroleum Review*, a leading trade journal of the oil industry, would disagree vehemently with Campbell. Not so. "In 1995 it all seemed pretty fantastic," Chris Skrebowski says. "I tried hard to prove him wrong. I have failed for nine years. I am now with him. In fact, I think he's a bit of an optimist."

Skrebowski, an oilman turned journalist, has compiled research in his own journal that shows why. In January 2004 he listed all oilfield "mega-projects" of more than 500 million barrels of oil.[122] Projects on this scale tend to have peak flows of around 100,000 barrels a day and account for around 80 percent of world supply. Only three projects of this scale are due to come onstream in 2007 and another three in 2008. Between 2003 and early 2007, in total, some 8 million barrels of new capacity is expected to come onstream. This "mini glut", as Skrebowski calls it – somewhat tongue in cheek – should be sufficient to offset global production declines from existing fields of about 3–4 million barrels a day over that period and projected demand growth of around 3 million barrels a day. But beyond? "There are not enough large-scale projects in the development pipeline right now to offset declining production in mature areas and meet global demand growth beyond 2007," Skrebowski says. From 2007, the volumes of new production will likely fall short of the combined need to replace lost capacity from depleting older fields and satisfy continued growth in world demand. His compilation shows that, on average, it takes six years from first discovery for a mega-project to start producing oil, leading to the conclusion that any new project approved today would be unlikely to come onstream until the end of the decade.

Skrebowski's 2004 study listed twenty-three other projects that could potentially be developed some time in the future. All but two of these are in Russia and the Middle East, as might be expected given the views of my senior source in BP cited earlier. But Skrebowski concluded that, due to a range of political, legal and technical uncertainties, none is likely to add new supplies to the market before the end of this decade. In April 2005 he updated the study, including fields that are smaller than 100,000 barrels per day but have the potential to grow beyond that, making a total list of seventy-three mega-projects due onstream by 2012. The principal change from the version published in 2004 is the number of projects where the date on which they are due to come onstream has slipped. "It remains true that few projects are listed beyond 2007/2008," Skrebowski observes.[123]

There is another complication. Almost all of the projects listed by *Petroleum Review* involve offshore oilfields. Since the infrastructure and operating costs of offshore projects are much higher than onshore projects, they are usually developed so that peak flows are achieved quickly – within about a year of start-up – and maintained for as long as possible. Offshore fields deplete more rapidly as a result.

This is a barely credible situation. Skrebowski is not easily able to spell out and explore the implications in the flagship oil industry journal which he edits, if he wants to keep his job. So I will do so. There will be nowhere near enough oil coming onstream to meet the combined forces of depletion and demand between 2008 and 2012. If there were such oil available, it would be from projects we would know about today, oil companies liking to boast to their shareholders about every sizeable discovery as they do. Given the inevitable lag time from discovery to production, there is now no way to plug that gap. There is worse. People in the oil industry must know this. They should be alerting governments and consumers to the inevitability of an energy crunch, and they aren't.

In July 2004, Campbell and Skrebowski tried to carry their warning jointly to the UK parliament. In the Thatcher Room they delivered a seminar to a pitifully thin audience, including only three members of parliament and a handful of researchers. I sat there listening to it with as surreal a feeling as I have ever experienced in all my years working on energy. Over the course of a decade at and around the climate negotiations, I have rarely been able to claim that the global warming problem is not reaching the ears it needs to. The same can manifestly not be said about the oil-depletion problem. This is the starting point for any analysis of how serious the problem is. How can evidence so compelling go almost unheard in one of the world's centres of government, even with a suspiciously high oil price at the time and so much obvious oil-related trouble brewing in the Middle East?

Having built their cases, the two spelt out the consequences of the early topping point. "The perception of looming decline may be worse than the decline itself," Campbell said. "There will be panic. The market over-reacts to even small imbalances. Prices are set to soar in the absence of spare capacity until demand is cut by recessions. We will enter a volatile epoch of price shocks and recessions in increasingly vicious circles. A stock-market crash is inevitable."

"If the economic recovery continues," Skrebowski added, "supply will get very tight from 2008 or 2009. Prices will soar. There is very little time and lots of heads are in the sand."

Matthew Simmons

Matthew Simmons, whose credentials within the industry need no repeating, views the crisis in a broader context. "Oil markets are headed towards a brick wall," he says. But they are not alone. Simmons stresses that oil depletion needs to be looked at in the context of other contemporary global energy problems. Most energy sources are separate "hubs", he points out. Oil is 70 percent transportation fuel. Natural gas is used primarily for heat and electricity. Coal and nuclear create electricity. All these hubs have their own problems. All have fixed capacity and long lead times to expand supply, with substitution options occupying pitifully small markets, as things stand. Gas prices have been going up as supplies near the main markets are depleting faster than expected. Because so much gas is burned for electricity, electricity bills have been going up too. Meanwhile, the electricity infrastructure in many countries is ageing fast, with huge capital expenditures needed on grids. For more than a decade, Simmons says, energy experts have immersed themselves in conventional wisdom which is wrong. He points to a cover of the *Economist* magazine from March 1999 as an illustration. The picture shows oil-soaked roughnecks struggling to plug a gushing well. "Drowning in oil", reads the headline. Energy-demand growth was supposed to be slow in the years after that article. Technology was supposed to reduce the cost of supply. The future cost of energy was supposed to fall. "All dead wrong," says Simmons. The oil crisis will hit, and be amplified by the problems in other areas of energy. "Is this a bad dream?" he asks. "Why did we wait so long?"[124]

Other early toppers

Chris Skrebowski is not the only oil-industry high-flyer who has tried to take on Colin Campbell and prove him wrong, only to end up converted. Richard Hardman, former Chief Executive of Amerada Hess, set out on his retirement a few years ago to use his spare time to assemble the arguments that would undermine Campbell's case. He ended up concluding they didn't exist. Hardman is now endeavouring to organize a high-level conference on oil depletion and other energy problems at the Royal Society. He hopes it will prove to be a "wake-up call" for the British establishment.

Roger Bentley, formerly of Imperial Oil in Canada, is another who has tried to blow the whistle in the corridors of power. Over a period of eight years he has lobbied the Department of Trade and Industry, BP, Shell, the Royal Commission on Environmental Pollution and others. "It's so frustrating," he told me, summarizing the experience. "Economists you can't convert. Engineers you have to tell and tell again, then they might get it." The UK government's Chief Scientific Adviser, Sir David King, told Bentley to come back when papers on oil depletion had been published in peer-reviewed journals such as *Nature*.

Another late convert is Roger Booth, who spent his whole professional life in Shell, most recently as head of their Renewables Programme. "I have reached that cynical stage in life that means I am concerned that when the peak does hit in the next few years, the reaction will be totally irrational," he told me. "A crash of 1929 proportions is not improbable. As the Chinese proverb states, 'May you live in interesting times.'"

A depression like the 1930s? How can we have let this happen? These are questions that many people will soon be asking. I examine how we got into the whole mess, and how we can escape or limit the damage, in Part Two of the book. In this chapter, I am still cataloguing the problems and assessing their magnitude.

BP's Francis Harper, the man who until recently co-ordinated their reserves data, won't be tied to a date, but certainly does not think the topping point lies far off in the 2020s and 2030s, as he made clear at the Energy Institute conference in November 2004. "My views are my own and not necessarily those of the company," he said at the outset. "The world is not facing a crisis, but neither should it be complacent." After the non-OPEC countries have gone over their topping point, the ability to close the gap between supply and rising global demand will depend on the larger OPEC producers. Harper's bottom line is coded. "The extent to which they can increase their production to both compensate for non-OPEC decline and satisfy a requirement for global growth will depend on economic and political decisions taken within the next ten, or even five, years."[125]

Therein lies the next issue, however.

IS THE INDUSTRY INVESTING ENOUGH TO FIND AND DELIVER OIL EVEN IF ENOUGH OF IT EXISTS?

Let us suppose for a moment that the late toppers are correct. The topping point, as defined by reserves available in principle, is off in the 2020s or 2030s, and we can look forward to growing supplies of relatively cheap oil for a decade or more. There is another aspect of the problem: whether or not the production capacity is sufficient. Oil-industry analyst Michael Smith, who took his PhD in geology just after me – sitting in the same chair as I did in the research lab – is an expert in this subject. He has spent most of his vocational life as an oil-industry geologist working around the world, particularly in the Middle East. "Reserves are largely irrelevant to the peak," he says. "Production capacity is the important thing – how quickly you can get it out. It is an engineering problem, not a geological problem." Of the eleven countries in the Middle East, only five are significant oil producers: Iran, Iraq, Kuwait, Saudi Arabia and the United Arab Emirates, known sometimes as the Middle East 5. They produce around 20 million barrels a day today, a quarter of the global total. If global demand rises at the average rate of the last thirty years, 1.5 percent per year, these five countries will have to meet around two-thirds of the demand, Smith calculates. Let us assume they can do what they say they can, no more, no less. Where does that leave us? Saudi Arabia says it can lift production from 9.5 million barrels per day today to 12 by 2016 and 15 beyond that. This despite 50 percent of the oil coming from the Ghawar field, where a water cut is already reported. Smith sums all the reported capacities in the Middle East 5 and finds that if the rate of demand growth continues at 1.5 percent they will fail to meet global demand by as soon as 2011. If it rises to 2.5 percent the demand gap appears in 2008. If it is 3.5 percent – the rates in China and the US of late – the gap is already here. "What's more," Smith adds, referring back wryly to the starting assumption, "I do not truly believe the claims of the Middle East 5. In fact although I don't believe Saudi and Iranian claims in particular, I think their politicians do believe them. I don't think there is a conspiracy, more a division of labour such that no one knows the whole story, each part of which has wide error bars. The summed result is inevitably the most positive conclusion which goes to the politicians. I've seen this in all the oil companies I have worked for."[126] At the November 2004 conference on oil depletion at the Energy Institute, Michael Smith showed a slide at the end of his presentation that gave a pictorial summary of his views. It showed a group of firemen posing for the camera outside a burning house.[127]

The investment bank Goldman Sachs drew attention to the problem of access to oil on a global scale in a much-quoted 2004 report. "The industry is not running out of oil – reserves are large and continue to grow," it asserts – though failing to offer evidence of this analysis. "What the industry is running out of is the ability to access this oil." Two decades of chronic underinvestment in the 1980s and 1990s are responsible. During this time the industry has been feasting on reserves discovered in the 1960s and earlier with infrastructure capitalized in the 1970s, after the first oil shock. Global oil demand is now closing fast on tanker capacity and refining capacity. The peak year for tanker capacity was way back in 1981. So too was the peak for refinery capacity. Global rig counts also peaked that year.

So, how much new investment is needed to fix the shortfall? Over the next ten years, assuming oil demand increases as commonly projected, fully $2.4 trillion will need to be spent, according to Goldman Sachs. This is nearly triple the level of capital investment by the oil industry in the 1990s. And if it isn't spent? "If the core infrastructure does not improve, energy crises are likely to become progressively more frequent, more severe and more disruptive of economic activity," the investment bank concludes.[128]

Stated simply, it seems that even if an early topping point doesn't hit us, the results of two decades of negligence in investment in infrastructure and exploration will. You need to read between the lines of the Goldman Sachs report to smell the level of anguish about this.

There is something else. At a briefing of analysts in 2004, Shell CEO Jeroen van der Veer listed another bullet point in the challenges he and other industry bosses face in delivering the goods. People. The average age of oil-industry personnel, including in the specialist service industries, is a staggering forty-nine. As one of my old colleagues from the oil days put it to me, what do you expect after decades of a ruthless hire-and-fire approach on top of all the environmental downsides? Legions of brilliant young graduates just dying to join the industry? The result of the hiring and firing, including of well-qualified staff, has been the concentration of an awful lot of experience and knowledge in a decreasing number of increasingly ageing hands.

THE GAUNTLET OF GEOPOLITICAL RISK EVEN IF INVESTMENT IS FORTHCOMING

Even where substantial money has been invested, a further list of serious unresolved problems can often be quickly summoned up. Oil in the Caspian is central to every scenario that envisages oil supply meeting demand off into the 2020s. The oil industry has long regarded the Baku–Ceyhan pipeline from Azerbaijan to Turkey as essential if it is to get Caspian oil out to market without the need to go through Chechnya and Russia. By the time this pipeline begins to shift oil as planned in 2005, it will have cost $4 billion, almost three-quarters of that in the form of bank loans. The problems for this pipeline begin with reports of its construction standard. Four whistleblowers recently told a UK national newspaper that the pipeline was failing all international construction standards, including installation of inadequately welded pipe before it had even been inspected. It passes through a major earthquake zone. Turkey has had seventeen major shocks in the past eighty years, and the pipeline is supposed to last for forty years.[129]

At the time the pipeline was conceived, industry reports talked of several hundreds of billions of barrels in the Caspian region. Now estimates of around 50 billion barrels, about the same as the North Sea, are more common.[130] After the discovery of the last of the super-giants, the Kashagan field in 1990, there was a burst of predictable interest in Kazakhstan. But now, in terrain where individual wells cost $1 billion to drill, in conditions where only foreign companies have the know-how and technology to drill, the Kazakh government has introduced new legislation that makes investment unattractive. As an ExxonMobil executive told *Petroleum Review*, "… the jury is still out on whether all these obstacles will delay Kazakhstan's production".[131]

This example of a real-world current problem for the oil industry raises the subject of the interplay between the early topping point and oil geopolitics. As the world's number one consumer, the United States will have much to say about how the crisis – whether of early depletion or inadequate infrastructure and investment, or both – plays out. The geopolitics of American oil dependency is well summarized by Michael Klare in his recent *Blood and Oil*.[132] He sees four key trends in US energy behaviour: more imports, increasingly unstable and unfriendly suppliers, escalating risk of anti-American violence and rising competition for diminishing supplies. Imports we have talked about above. Increasingly unfriendly suppliers and escalating anti-American violence are linked. The point here is that the US can have relationships with governments in unstable countries if it chooses the path of oil dependency, but not easily with their populations. Terrorism can be expected to grow with every American act interpretable as imperialistic in the Middle East and Central Asia. The Iraq-to-Turkey pipeline illustrates the problem perfectly. It suffered near daily attacks in 2003.

As for competition over diminishing supplies, therein lies the stuff of nightmares. The Pentagon established a Central Command in 1983, one of five unified commands around the world, with the clear task of protecting the global flow of petroleum. "Slowly but surely," Michael Klare concludes, "the US military is being converted into a global oil-protection service." At $30 a barrel, the total bill for imported oil – now more than half the US daily consumption and rising fast – should reach $3.5 trillion over the next twenty-five years, and this does not include the Pentagon's overhead. Beyond the Middle East 5, the Bush strategy of supplier diversification will look to eight main sources, which Klare calls the Alternative 8: Mexico, Venezuela, Colombia, Russia, Azerbaijan, Kazakhstan, Nigeria and Angola. These countries and their oil operations are characterized by one or more of the following attributes: corruption, organized crime, civil war (in five of them in recent years), political turmoil short of civil war, and ruthless dictators. The US military is being forced into deeper relationships with such regimes, including joint military exercises.

The bottom line for Klare is this. "Any eruption of ethnic or political violence in these areas could do more than entrap our forces there. It could lead to a deadly confrontation between the world's military powers." Because obviously, in a world as enduringly addicted to oil as ours is, others are going to be looking for their own supplies. Russia and China will be among them. As one global-security analyst recently put it, "I am afraid that over the years we will see China become more involved in Middle East politics. And they will want to have access to oil by cutting deals with corrupt dictatorships in the region, and perhaps providing components of weapons of mass destruction, ballistic missiles and other things they have been involved with, and that could definitely put them on a collision course with the United States."[133] Oil dependency could yet prove to be the route to a Third World War. The stress associated with an unforeseen early topping point surely makes that horrific prospect more, not less, likely.

On this unpleasant note, I come to the end of my analysis of the imminent crisis that is the early oil topping point. I have not yet considered potential solutions or options for escape. This is because the problem conflates with another, which we must consider before tackling solutions and escapes in the round. This one brings us full-square back to Matthew Simmons's insistence that any solutions to an early oil peak must be viewed in the context of global energy problems as a whole.

In introducing the issue of global warming, and Part Two of the book, I think of a foretaste of the kind of impacts the super-high energy prices of a global energy crisis will entail, a hint of the extremes to be expected in a world in climatic meltdown, all in the context of the realities of the energy regime we have today, rather than the one we might like to have tomorrow.

23 December 2003, Waterloo, London

It is as cold in London now as it was hot just four months ago. An elderly couple have frozen to death in the house they have lived in for more than sixty years. Their gas supply has been disconnected these last two weeks because they couldn't afford a small bill. The coroner has recorded a verdict of "death by natural causes".[134] I try to digest the fact that I am now living in a land where it is not only legal but natural for energy companies to abandon people to death by hypothermia in pursuit of unpaid bills. A whole new dimension in gross margin, I think to myself. In Finland, a somewhat colder country than Britain, the toll of people freezing to death in their own homes in a typical year is precisely zero. The annual toll from hypothermia deaths in British homes, on average, approaches 50,000.[135] Tens of thousands of grans and grandads dying who wouldn't have died if they lived in the land of Father Christmas with a little insulation in their homes! All we Brits would have to do to stop this particular form of carnage in our society is figure out how to make our homes as energy efficient as the Finns do. But we don't bother. We permit our bog-standard buildings to be leaky sieves pissing warmth and greenhouse gases into the atmosphere as though the lives of nice old codgers, the looming energy crisis and the average temperature at the surface of the earth have no value whatsoever.

It seems only yesterday that we were as worried about heat as we are now about cold. Thousands of people, mostly the old, were dying in the hottest summer on record. I think of a day late in August. Even before the lights went out, there was a certain atmosphere, a lingering distillation of the strange things that happen in heatwaves. You know what I mean. You walk into the office and your first thought is to wonder who it is that is inadequately familiar with deodorant. Faint snarls begin to appear on the faces of people you never normally exchange a cross word with. Workaholics start to lust after cold lager at 3 o'clock in the afternoon.

It had been breathlessly hot in London for weeks. On 10 August, the thermometer topped 100° Fahrenheit for the first time ever in the UK. Railway tracks buckled, trapping thousands in saunas that had once been mere carriages. The roads proved no better a means of getting around, because they were melting.

The mercury kept right up there, day after day, for three weeks. At London Zoo, they began feeding ice creams of frozen blood to the panting lions. Our office had no air conditioning, and by mid-afternoon you felt that your brain was turning into marshmallow. Knowing we were breaking health and safety rules by the chapter, I told everyone they could go home if they wanted, and ran out to buy ice creams to encourage them to stay. "Frozen blood," I announced. "Come and get it!" The good troops sat pink-faced at their computers, rummaging among their marshmallow brains for coherent thoughts, trying not to get ratty with customers, much less each other. Not long after 5 o'clock on this particular day, about a dozen of us could hold out no longer against our most consistently coherent thought. So there we stood, in our inappropriately named local, The Fire Station, shoulder to shoulder with an already rowdy host of people, drinking cold lager with that certain atmosphere in the air.

That's when the lights went out.

We looked at each other in the sweaty gloom. We live in such strange times, and a power cut is a rare event, so the speculation began at once. Perhaps it was the terrorist attack on London that our Prime Minister had been encouraging us all year to think inevitable? We sniffed the air. "Sod it," someone said, and rushed to get another round in. He returned disappointed. No electricity. The pumps were down. Mobile phone calls were made. None got through. Scouts left to check the outside world. All the lights were out in the neighbourhood.

Could the cut be nationwide? We had context. Two weeks earlier North America had suffered its worst-ever power cut. Twenty-one power plants had gone down, leaving 60 million people in New York, Detroit, Toronto and dozens of other cities without electricity all night, trapping commuters in stifling trains and lifts, forcing thousands to sleep on the streets. Airlines had to cancel flights all across the world. We had seen the TV pictures: the thousands tramping across the Brooklyn Bridge against a blacked-out Manhattan; the Wall Street bankers sleeping on park benches.

Scenes as though from the collapse of civilization.

A day later it was over, with little harm, we were assured, beyond the half-billion dollars in lost business clipped from the American balance sheet. But unanswered questions lingered. Last time a major power cut happened on that scale in America, in November 1965, we were surprised to learn that the failure had been traced to a single faulty relay in Ontario. This time, the US and Canada were still bickering about the nationality, much less the exact location, of the fault responsible. (It turned out to be two sagging power lines in Ohio.) Surprise turned to amazement when a former US energy secretary told the media that the American electricity grid was of no better quality than those in the Third World. A UK electricity expert immediately leapt into the headlines by predicting that there was a one in five chance of the clapped-out British grid going down in the same way before Christmas.

Maybe the French were responsible, someone ventured. I'm ashamed to say that this is a common thought in the land of Foot and Mouth, Mad Cow Disease and death-trap railways. Earlier in the month a quarter of French nuclear power stations had become so hot – for the heatwave was Europe-wide – that they had to be shut down. Seventy-five percent of French electricity is generated by nukes, and the British national grid taps some of it across the Channel. Could the insane prospect of a France too hot to cool its nukes have anything to do with the tragic fact that our pub could no longer pump lager? Yes, damnit, maybe the French were to blame!

The minutes ticked by without the smell of sarin or the sound of air-raid sirens. We speculated on.

Half an hour later, the lights came back. A cheer rolled around the pub, and a scrum descended on the bar. Being next to Waterloo Station, the place was now bursting. Recognizing rightly that any power cut would knock out the national transport infrastructure beyond immediate or even mid-term prospect of repair, city workers were settling in here and in other pubs all across the capital for a night of partying. Our August power cut had not been anywhere near as protracted as America's, but bankers would still be sleeping on park benches that night.

II

OIL DEPLETION MEETS GLOBAL WARMING

CHAPTER 5

Depletion in the greenhouse

I laid out the arguments in Part One of the book on the assumption that we can afford to go on burning oil and gas for as long as we can find and pump them. Most economists and financial analysts operate in a culture that assumes we can. But they are wrong. We can't. The reason is that inadequately named phenomenon, global warming. If we carry on burning oil, gas and coal at the rates we're currently burning them, global warming is also quite capable of booting us into the next depression or worse, while wrecking ecosystems along the way. In the UK, Sir David King, the government's current Chief Scientific Adviser, and Sir John Houghton, the past head of the Meteorological Office, have both said that global warming is now a bigger threat than weapons of mass destruction.[136] This is an increasingly common view, especially in Europe.

Let me first explain why such views are justified. Thereafter I will give an overall threat assessment of global warming – Big Oversight One in The Story of the Blue Pearl – and how it conflates with oil depletion, Big Oversight Two.

GLOBAL WARMING AS A WEAPON OF MASS DESTRUCTION

Imagine that back in the Cold War the CIA discovered that the Soviet Union had secretly developed, and deployed, two dreadful new weapons of mass destruction. One was a form of economic time bomb able to wreck Western economies – indeed bring down capitalism itself. The other was a biological weapon able to wipe out an entire ecosystem not found in the USSR – an economically vital one that many countries quite literally base their economies on. The response would be easy to imagine in the light of world events today. The B-52s would be revving up, laden with nuclear missiles, at the end of every US air force runway. US troops would be packing their ammunition in all seven hundred American military bases around the world. NATO countries would be queuing up to support their ally in her enraged insistence that the Soviets back down. The Cuban missile crisis would seem like a trifling quibble in comparison.

Why is it that we take such a different approach to environmental security? For both those "weapons of mass destruction" exist here and now on Planet Earth, though they have not been invented by any superpower. What is more, they are beginning to take effect. The economic one involves the threat global warming poses to the survival of the trillion-dollar global insurance industry, and knock-on impacts on capital markets. The biological one involves the real and present danger global warming poses to the survival of coral reefs.

9 February 2000, City of London
I am sitting in a room with forty fund managers who control £250 billion of assets. By way of comparison, the FTSE all-share index today has a combined value of some £1,500 billion. In the two hours I have spent in London's golden Square Mile, well over a trillion dollars has

changed hands in the global capital markets. Most of it involves pension funds and mutual funds – run by the kind of people I am sitting with – whose $9 trillion combined worth dominates the $20 trillion world economy. Much of the rest involves banks and insurance companies.

The monied ones are gathered this morning to listen to analyses of a threat to their billions and trillions. Fund managers often get together for such seminars, but never before have so many been tempted from their offices by today's subject. I have finished my own presentation and am sitting among the pin-striped suits listening to another speaker, Dr Julian Salt, of the Loss Prevention Council, an insurance industry think-tank.

"The insurance system that operates today is fatally flawed," he begins. "The system relies on the collection of claims data from the past to underwrite in the present for the future's unknown claims." This is fine in a system that never changes, Salt explains, but unsustainable in a world with a changing climate and the potential for unmanageable claims. The insurance industry has annual income in excess of $2 trillion, fully 10 percent of the global economy. The entire shooting match is in dire peril because of global warming. The insurance industry itself is at risk of bankruptcy. That could create a domino effect. Pensions are at risk. Banks could fall. Indeed, global capital markets could collapse.

He doesn't need to spell out the obvious corollary. If this happens, the value of the fund managers' investments will be blown out of the water.

The only answer, Salt concludes, is full-scale retreat from the fossil-fuel burning that poses the risk of unmanageable claims in a changing climate.

"We need to decarbonize industry, and energy production," he says.

This has been a belief of mine for many years. I have argued the case to insurers and bankers since 1993. I have rarely been able to pitch to fund managers, though, and never with so much money in the room as this day. Salt is making a better job of it than I did earlier in the conference, and not just because he comes from their world.

Salt delivers his talk deadpan. It will all become ever more clear to people, he says, with each huge insurance loss, and every near miss. By this he means near misses on cities by superstorms, drought-related wildfires and floods, all of which become more frequent as the planetary thermostat rises. Just a couple of these – a typhoon hit on Tokyo, say, or a wildfire making it into Los Angeles – hold the potential to drain the global industry's entire property-catastrophe reserves. The industry holds something in excess of $300 billion for all natural-catastrophe losses in a year. That could in principle be wiped out overnight. No major climatic catastrophe has ever hit a city, and the insurance industry is behaving as though it never will.

Salt offers a simple and clear conclusion: "The message for future investors should be 'Buy silicon, sell carbon'."

There are a few problems with this maxim as things stand, he admits. "The market capitalization of conventional fossil stocks is too large. The alternative non-fossil stock capitalization is too small. To change things round, we need a hundred Microsofts and ten Kyoto Protocols." He means Microsofts dealing in alternative energy, not software, and treaties ten times stronger than the protocol negotiated by most of the world's governments in 1997, which will barely begin the process of cutting greenhouse-gas emissions, even when it is eventually ratified by enough governments to come into force.

Solar Microsofts and climate treaties with teeth could well emerge,

Salt thinks, in the years ahead. History isn't destiny. Investment tides turn. Political pressures could turn on their heads. If we want to save the planet, he warns, they will have to, and quickly.

After the meeting, I collar Julian Salt over the buffet lunch. What do his insurance-industry colleagues say when he delivers this kind of analysis in-house? I ask.

They listen, he replies with a grim smile. Then they generally go to the bar.

I have already toured the fund managers canvassing reactions. For the most part, they have never heard the argument before. Some have thrown an attempted rebuttal back at me that I am all too familiar with. There is a lot of uncertainty about the science behind global warming, isn't there? How does one know who to believe?

When are they going wake up? I ask Salt. When is the tide going to turn, do you think? Isn't there enough evidence already? I bet these guys will go right back to their offices and keep investing squillions in oil, gas and coal.

The financial services industry has a great big capacity for denial, Salt replies.

This I am already overly familiar with. But how big, exactly? That is what I want to know. At some point, surely, the march of evidence will tip the dysfunctional capital markets into behaviour designed to promote their own survival, as opposed to suicide.

The new century is only six weeks old, but already there is plenty of writing on the wall, for those with eyes to see. The latest taste of what awaits in a continuingly fossil-fuel-addicted world hit France on the eve of the new century: the worst storm in a thousand years. Two nights of vicious winds swept across the country on 26 and 28 December, clocking up the most costly catastrophe in Europe's history. More than half the roofs in Paris were damaged, 300 million trees came down: 3 percent of all France's woods. The electricity network suffered the greatest damage ever seen in an industrialized country short of wartime: 2 million people were left without power immediately after the storm, and many remained so for a long time. Thousands of troops were out trying to get the country back on its feet.

CNN has woken up too, it seems. Throughout the 1990s, the global news network's coverage of global warming routinely embraced what producers thought of as balance. This involved pairing any interviewee urging greenhouse-gas emissions cuts because of the threat of global warming with a contrarian – usually from a hard-core pool of half a dozen, often in the pay of the oil industry – smoothly asserting that there is no threat, or only a small one. In the first week of the new century, when CNN invited me to their London studio to appear live on an hour-long round-the-world programme on global warming, I anticipated the kind of high-blood-pressure bout with flat-earthers that I had taken part in before on the network. But I was to be surprised. The programme opened with the presenter wondering, in the kind of drama-edged voice usually reserved for America's latest firearms atrocity, whether – given the dire evidence emerging about the threat of global warming – the next century might in fact be the last for humankind.

What has changed CNN's mind, it seems, was the announcement by American scientists that over the last forty years the Arctic ice cap has thinned by around 40 percent. This evidence, which featured large in the programme, came from American nuclear submarines playing Cold War hide and seek with Russian counterparts. The Pentagon has been sitting on the knowledge for years. National security, it seems, applies in the collective mind of the state only to the extent that nuclear weapons might affect national economies, not the planet on which we live.

Laying waste to economies and ecosystems alike

Julian Salt was not saying anything particularly new at that investors' meeting. Insurance leaders have been warning since the early 1990s that, if we carry on putting heat-trapping gases into the atmosphere by the billions of tonnes each year, their industry could be bankrupted by global warming. The global reserve the industry keeps for property-catastrophe insurance payouts could be wiped out either by one or two "mega-cats", as the insurers call big hits on cities, or by a machine-gun fire of smaller catastrophes. The trend is there to see already: losses from natural catastrophes have been rising at around 10 percent per year in constant dollars for four decades. Major insurance companies such as the two biggest reinsurers in the world, Munich Re and Swiss Re, do not use conditional language when they describe it in their annual reports. The trend is real and in large part ascribable to global warming.

If it continues, by around 2060 we will be destroying wealth as fast as we can create it, even assuming no interim economic dislocations.[137] As long ago as 1993 I listened to a director of Lloyd's of London warn that enhancing the greenhouse effect could bankrupt not just Lloyd's but the entire global insurance industry. By 1995, having focused on researching this remarkable problem, I had also heard industry leaders warn of a greenhouse-triggered global insurance crash in Tokyo, New York, Munich, Zurich and Bermuda – most of the world's major insurance centres.[138] In 1997, the world's largest reinsurance company went further. Munich Re wrote in a report on the 1996 catastrophe record that "... according to current estimates, the possible extent of losses caused by extreme natural catastrophes in the one of the world's metropolises or industrial centres would be so great as to cause the collapse of entire countries' economic systems and could even bring about the collapse of the world's financial markets."[139] Swiss Re's 2004 report says "... there is a danger that human intervention will accelerate and intensify natural climate changes to such a point that it will become impossible to adapt our socio-economic systems in time."[140]

These warnings, it should be emphasized, come not from environmentalists but from two of the biggest corporations in the world.

The warnings about mass extinction in coral reefs come similarly from those who should know best: a majority of the biologists who study them. If water temperatures rise just a degree, for as short a period as a few weeks, corals begin to bleach. This whitening effect is related to a loss of ability to feed, and it can kill them if the water remains too warm. Coral biologists reported worrying levels of bleaching as far back as the early 1990s. Now this sickening phenomenon can be found in every ocean where corals exist. Some experts are worrying openly that corals are in a decline that could lead to their complete extinction as a result of global warming within as little as thirty years.[141]

The threats in summary: a baker's dozen

These are far from the only threats posed by global warming. More than ten years ago a body of experts warned that unless we make deep cuts in the burning of the main source of emissions – oil, gas and coal – the effects of global warming could ultimately be "second only to nuclear war".[142] In my book *The Carbon War* I summarized twelve main reasons for concern if greenhouse-gas emissions remain anywhere near today's levels in the decades to come. Distilled from a vast body of research carried out by climate scientists around the world, as presented in three reports by the Intergovernmental Panel on Climate Change (IPCC),[143] they are in brief:

- *The degree of warming.* Projected levels are too high for many ecosystems to tolerate, certainly high enough to trigger non-linear reactions such as the switching off of the Gulf Stream.
- *The rate of warming.* Projected rates of warming, assuming no efforts to cut greenhouse-gas emissions, are too fast for many ecosystems to adapt to. For example, ecologists envisage whole forest types disappearing.
- *Biodiversity loss.* Corals are just one example. Recent climate modelling at the UK Meteorological Office has shown that the tropical rainforests – the most diverse ecosystem – are particularly at risk over a longer timeframe.
- *Sea-level rise.* In a world where much infrastructure and many of the mega-cities are on coastal plains, even small increases from the thermal expansion of warmer sea water are a problem. Yet melting ice, particularly in Greenland and Antarctica, can cause rises measured in metres over longer periods.
- *The threat to the insurance industry and capital markets.* Imagine, for example, the vulnerability of the $2 trillion of insured assets on the coast of a single state, Florida, and the threat this poses to an industry that keeps less than half a billion dollars in reserve to cover all catastrophe losses everywhere in the world in any one year.
- *The threat to food supplies.* Increased flooding, drought, heat stress and proliferating pests are all individually big problems for food supply. Worse yet are the synergies between these and other factors.

- *The threat to water supplies.* In a world already rapidly mining its aquifers, the increased drought projected for the warming world will significantly compound problems.

- *The threat to human health.* Of particular concern is the vertical and horizontal spread of areas where disease-bearing insects, notably the mosquito, can thrive.

- *The increased risk of conflict.* Many major river basins run through two or more countries. Several have experienced water-related conflict even without the added impacts of global warming.

- *The threat of societal instability.* Hundreds of millions of environmental refugees can be expected, for example.

- *The probability of amplifying feedbacks.* Heat-stimulated emissions from sources such as melting permafrost, drying soils, dying forests, stratifying oceans and melting methane hydrates hold the potential to make global warming even worse than the already-dreadful "best guess" estimates of the IPCC. Feedbacks with a suppressing effect, such as increased cloud reflectivity, hold much less potential to dampen warming.

- *The danger of a runaway effect.* The worst-case analysis is the risk that stimulated emissions via amplifying feedbacks might come to outrun emissions at source, taking us beyond a point of irreversibility. I return to this horrible possibility later.

HOW MUCH WARMING AND HOW MUCH DANGER?

As greenhouse contrarians will be quick to tell you, there are many uncertainties. But there are also many worrying aspects of the problem that we can be sure of. We know that greenhouse gases trap heat. We know that carbon dioxide is the main greenhouse gas, and that oil burning is its number one source. We know past concentrations accurately, going back 750,000 years, from measurements of air trapped in the Antarctic and Greenland ice cores. We can measure ancient air temperatures in a proxy way in these cores using the two different isotopes of oxygen.[144] The correspondence between the carbon-dioxide level and air temperature is so close it is hardly worth plotting the two apart. In Figure 6, the concentrations from the ice cores are plotted from 400,000 years ago to the beginning of modern instrumental records in the nineteenth century.[145]

The regular ups and downs over the thousands of years represent ice ages, in which carbon-dioxide concentrations hovered around 200 parts per million (ppm), and the warmer interglacial periods between them,

Figure 6: Carbon-dioxide concentrations, past, present and future (where future = unless we cut oil, gas and coal burning deeply)

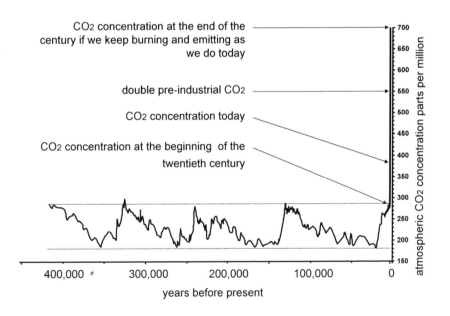

when concentrations were around 260–270 ppm. Note the rise in concentrations before fossil-fuel burning began in earnest, to 380 ppm-plus at the time of writing. If we keep burning fossil fuels at the escalating rate we do today the atmospheric concentration will approach 700 ppm by the end of the century in which we live. Quite a spike.

Zooming in and looking at the last 1,000 years and the next 100, let us see how the carbon-dioxide and other rising greenhouse-gas concentrations have translated, and are forecast to translate, into average global temperature increases. Figure 7 shows data from the IPCC Third Scientific Assessment.[146] For the last thousand years or so, the world has been in an interglacial period. The atmospheric carbon-dioxide concentration stood at around 280 ppm when the industrial age began, resulting in a rather steady global average temperature, depicted in Figure 7 by a moving average based on proxy data from ice cores plus coral and tree-growth rings in the northern hemisphere, including uncertainty ranges in the data shown in grey. Entering the period of global instrumental observations, which began in 1861, we see the global average temperature rising 0.8°C by 2004, as recorded by weather stations around the world. The trend is clear. In fact, the ten warmest years have all been since 1990, including each year since 1997.[147]

Where is the global average temperature heading if we keep burning fossil fuels at or near the rate we do today? The projected temperature range shows the estimates from the UK Meteorological Office computer simulations of a range of emissions scenarios envisaging a world which remains, to differing degrees, fossil-fuel dependent. The spike of projected atmospheric carbon-dioxide concentrations in Figure 6 hikes the planetary thermostat by between 2.5°C and 5°C above pre-industrial. That means the global average temperature would rise from 14.8°C to as much as almost 20°C. Very obviously we are heading into completely

Figure 7: Average atmospheric temperatures, present and future
(where future = unless we cut oil, gas and coal burning deeply)

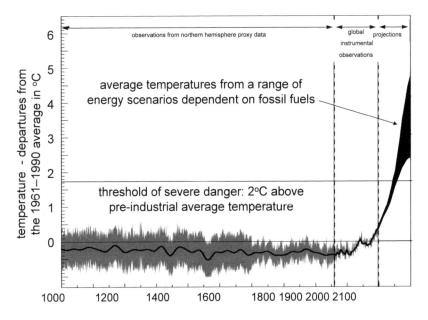

uncharted territory, even if the lower end of the range applies. These days you can find very few scientists who do not think this involves massive danger to the planet.

The danger threshold: near, even according to the European Union
Where might any threshold of unacceptable danger be? A common view is that we must not let the global average increase more than 2°C above the pre-industrial level. All fifteen then-members of the European Union accepted this as a policy target in 1996. As things stand, we crash through that ceiling of danger before or slightly after the middle of the century.

What exact level of atmospheric carbon-dioxide concentrations equates to 2°C cannot be known with certainty yet, because of the difficulty of calculating the extent to which amplifying feedbacks will speed warming. The Meteorological Office's latest best estimate for global average temperature rise following a doubling of atmospheric carbon dioxide from 280 ppm to 520 ppm is 3.5°C. The real figure could be as low as 2.4°C – still above the EU-defined danger threshold – but as high as 5.4°C.[148] As Sir David King put it in 2004, "I don't think we should be anywhere near 550 parts per million of carbon dioxide in our atmosphere."[149]

Analyses like these encouraged the UK government to announce that global warming would be one of its two grand themes when it assumed the chairmanship of the G8 group of industrialized countries in 2005. In January 2005, the UK released a report written by some of its leading researchers, along with opposite numbers in the US and Australia, and the Chairman of the IPCC. It concluded that the 2°C danger threshold would be inevitable once the atmospheric CO_2 concentration reached 400 ppm. This was the lowest an official report had ever gone on setting the danger threshold. The report warned that "… above the 2° level, the risks of abrupt, accelerated, or runaway climate change increase". Given that 400 ppm is only 20 ppm above the current atmospheric concentration, that meant that "the point of no return", as media articles on the report called it, was less than ten years away on the current global emissions trajectory.[150]

One of the drivers for such fears is that the rate of build-up of atmospheric carbon dioxide could well be accelerating. The best recordings of atmospheric carbon-dioxide concentration are made in Hawaii, where an observatory has been making detailed measurements for nearly half a century. For most of that time the concentration has risen an average of 1.84 ppm per year. In the last two years it has risen at 3 ppm annually. Scientists do not yet know whether this is a temporary rise, but it could very well be the effects of amplifying feedbacks kicking in. Top candidates would include an overall declining ability of increasingly stressed forests to take carbon dioxide out of the air during photosynthesis, and a decreasing ability of phytoplankton to do the same in the oceans. To say the least, researchers are worried. The record in 2005 and beyond will be closely watched.[151]

Most governmental agencies responsible for studying climate change, such the Meteorological Office in the UK and NASA in the US now profess that we are seeing the first impacts of global warming in extreme weather patterns around the world. In 2004 we saw the hottest ever summer in California with record wildfires as a result, atypical flash floods in the UK, an unusual battery of damaging hurricanes in Florida and the complete collapse of seabird colonies in Scotland as their food supply vanished in unusually hot waters. Whether or not such impacts can be ascribed definitively to the early emergence of global warming is not the most important issue, however. Far more important is the strong degree of certainty we can place on the far worse extremes to come if we are foolish enough to let greenhouse-gas concentrations go on building up in the atmosphere on anything like the business-as-usual trajectory shown in Figure 6. These extremes hold ample power to destroy economies. Whether we have as much oil and gas to burn as the late toppers would have us believe, or as little as the early toppers, we have plenty to tip us into global economic ruin as a result of climatic meltdown.

FEEDBACKS: SLEEPING GIANTS ON THE ROAD TO HORROR

Increasingly the danger feared most by climate scientists is the capacity of the climate system for abrupt changes and thresholds of irreversibility. Scientists at the Tyndall Centre, one of the UK's national climate research centres, have identified around a dozen so-called "tipping points" of particular danger. These they refer to as "sleeping giants". A major gathering of international scientists in February 2005 assessed these sleeping giants in a conference preparing for the G8 summit. The outcome was a shrill warning to the G8 leaders: the giants are in danger of being awoken imminently. As John Schellnhuber, Director of the Tyndall Centre, put it: "If we go beyond two degrees we will raise hell." The scientists endorsed the view that a mere 400 ppm of carbon dioxide could do this, and that such a level could arrive within just ten years.[152]

My personal top five concerns are as follows.

1. *Methane-hydrate destabilization*

    Eminent scientists now fear that the very same deposits of unconventional gas that some wish to mine in the growing scramble for energy can escalate global warming sub-stantially. I can think of no more potent example of the dysfunctional nature of the appliance of science in our modern world than this. Scientists at the United States Geological Survey calculate that there might be as much as 10,000 billion tonnes of carbon in methane-hydrate deposits.[153] They and other energy-industry researchers spend time wondering how to solve the problems of energy supply by mining these ice-like hydrate solids, as described in Chapter 3, and burning them. This despite knowing, presumably, that methane is many times more powerful than carbon dioxide as a greenhouse gas, molecule for molecule, and that much methane would inevitably escape into the atmosphere during any mining process, quite apart from the carbon-dioxide emissions from the burning. Meanwhile, climate scientists sit and worry that there is far too much carbon in existing gas and oil, never mind unconventional gas, to keep us below the danger threshold for carbon dioxide in the atmosphere. I return to that below.

2. *Soils cause land biosphere to turn from absorber to emitter*

Warming soils would emit more carbon dioxide than they do today. At the moment plants on land are a net reducer of carbon dioxide in the atmosphere, because of photosynthesis, which absorbs about a quarter of the carbon dioxide emitted by fossil-fuel burning. In a warming world decomposition of organic matter in soils would speed up, ultimately causing the land biosphere to switch states from sink to source for carbon. Some models suggest this could happen at an atmospheric carbon-dioxide concentration of between 400 ppm and 500 ppm, in other words between a decade and half a century from now.[154]

3. *"Ocean conveyor belt" shuts down*

The ocean current system in the North Atlantic is driven by a natural conveyor process: the sinking of dense, salty water off Greenland pulls warm water north from the Caribbean region. This surface current, the Gulf Stream, has a massive ameliorating impact on the climate of western Europe. The conveyor could switch off if enough ice melts in Greenland and the Arctic to stop the surface water from being salty enough to sink. It has done so in the past, most recently 12,000 years ago. In a rapidly warming world, western Europe could then become significantly colder.[155]

4. *Greenland ice cap melts*

Some 6 percent of all the fresh water on the planet, 2.6 million cubic kilometres, sits frozen in the 2-kilometre-thick ice cap atop Greenland. It can melt. If it did so, global average sea level would rise fully 7 metres, inundating coastal cities and plains where most of the world's population lives, and where most economic and agricultural activity is focused. Researchers now worry that it will do so in a slow but irreversible process if regional warming exceeds around 2.7°C.[156] The process might take more than a millennium to complete, but you can just imagine what the history books of 3004 (if they ever get written) would have to say about the governments and industrialists of 2004 if this comes to pass.[157]

5. *West Antarctic ice sheet slides into the sea*

The Greenland ice sheet is not the only way to lift global average sea level by metres. The West Antarctic ice sheet is more than 3 kilometres thick, and is pinned to the continent by ice shelves attached to the continental shelf. If these shelves melt from below in warming waters, the whole edifice could slide off into the oceans. If that happened, the seas would rise around 5 metres within a few hundred years, regardless of whether the ice melted or not.

At the Exeter G8 preparation conference in February 2005, the Director of the British Antarctic Survey, Chris Rapley, warned that "… we could be seeing the start of a runaway collapse of the ice sheet". The reason for such concern is that three of the biggest glaciers on the continent, around Pine Island, are losing a massive amount of ice: an estimated 250 cubic kilometres per year.

If this particular sleeping giant is triggered, it could easily trigger another. Loss of the West Antarctic ice sheet could destabilize the even bigger East Antarctic ice sheet, which could raise global sea levels 50 metres.

Other sleeping giants

The other potential sleeping giants pepper the world, geographically. In the Sahara, where more rain is forecast, dry dust currently blowing in prevailing winds to provide crucial nutrients to the Atlantic may no longer blow. Because the nutrients fertilize phytoplankton at the bottom of the food chain, oceanic biodiversity may collapse, with a net release of carbon into the atmosphere because of plankton loss.

In the Amazon, rainfall is forecast to drop, leading to the gradual death of the rainforest, massive biodiversity reductions, and the release of additional carbon into the atmosphere in the amplifying feedback mentioned earlier. The amount of carbon released by this one impact alone could, in a worst case, equal all the twentieth-century fossil-fuel emissions.

On the Tibetan plateau, which spans a quarter of China, melting of snow and ice could greatly reduce reflectivity of the planet (the "albedo" effect), meaning that less solar radiation is reflected back into space, with the effect that more heat is retained.

More than a third of the carbon dioxide humans emit ends up dissolving into the oceans as carbonic acid. Increasing oceanic acidity is already measurable, and an increase would have a devastating impact on oceanic life.

The main point is clear. Uncertainty there may be in the news, but none of the news is good. The worst-case analysis is that most or even all the spins of the wheel of fate that can go against us do go against us. That is why a growing number of scientists now talk about the dreadful worst-case prospect of a runaway effect.

THE INTERNATIONAL EFFORT TO RESPOND TO THE DANGER SO FAR

Concern began to nucleate in the mid-1980s, as scientists with some of the fastest computers in the world modelled future climate with realistic assumptions about greenhouse-gas emissions and found worrying levels of warming. In 1988 a withering drought hit the US midwest and giant forest fires swept the west of the country, partly destroying Yellowstone Park. The US Congress took notice. Governments, under the auspices of the United Nations Environment Programme and the World Meteorological Organization, put together an advisory body, the Inter-governmental Panel on Climate Change (IPCC), to study the problem and report back by 2000. The IPCC pulled in many hundreds of experts from government institutions, industry and academia to produce three assessments in one: on the science of the problem, the impacts and the solutions. I describe the history of what happened next in *The Carbon War*, which covers the years 1989 to 2000.[158] In summary, the IPCC's First Scientific Assessment Report, completed in May 1990, kicked off more than a decade of negotiations that are still under way. It concluded that greenhouse gases would certainly hike the global average temperature in the years ahead, and to a worrying degree.[159] At the World Climate Conference of November 1990, in Geneva, more than a hundred governments agreed to open negotiations in 1991 aiming to control the problem and intending to complete a binding treaty by the time of the June 1992 Earth Summit, in Rio de Janeiro. At that time, every industrialized country except the United States wanted legally binding targets and timetables for emissions reductions.

The Convention on Climate Change, 1992

They got the treaty, but minus the teeth. The Framework Convention on Climate Change, as signed by the world's leaders, had no targets and no timetables. George Bush Senior's negotiators, their many lobbyist allies from American oil, coal and auto companies, and the OPEC governments that generally did the diplomatic dirty work for them, ensured that.

What the Climate Convention did have, though, was a serious ultimate objective. It stated that the goal of the treaty was to stabilize atmospheric greenhouse-gas concentrations at levels that pose no danger to economies, food supply and ecosystems.[160] The IPCC's scientists had shown that, if the goal was to stabilize concentrations of greenhouse gases in the atmosphere, cuts in emissions of around 60 percent would be needed at some point. Stated another way, the treaty was an agreement at some point – short of danger – to replace most fossil-fuel burning with other forms of energy and energy efficiency. The argument has been raging ever since about where the level of danger lies. We have already discussed the European view of that. The US view is that it cannot be defined.

The Kyoto Protocol, 1997

Negotiations began over a follow-up process aiming to cut emissions. The Clinton Administration replaced the first Bush Administration and hopes rose, only to slowly deflate. The negotiations began to run out of steam between 1993 and 1995. They were reinvigorated in 1995 by completion of the IPCC's Second Scientific Assessment. The first had concluded that a "signal" of human enhancement couldn't be detected and probably wouldn't be distinguishable from natural climate variation for a decade or more. The second assessment reversed that view, saying that the first faint footprint was now visible in extremes of temperature. It also increased the estimates of warming from climate models.[161]

This new warning gave the negotiations enough momentum to attach a weak set of teeth to the convention. The Kyoto Protocol, completed amid emotional scenes in December 1997, is a legally binding document that commits the governments of industrialized countries to caps on emissions of a group of greenhouse gases including carbon dioxide by different agreed amounts from 1990 levels by or within a target window of 2008–2012. The individual country goals range from an 8 percent cut to a 10 percent increase on 1990 levels. Almost all involve reductions from present levels. The overall global reduction is 5.2 percent: a mere start towards 60 percent cuts, but an important one.

Since Kyoto

Of course, the protocol had to be ratified before it came into force. A critical mass of governments, representing 55 percent of all emissions, was needed to do this, with America and Russia being key to the deal. Clinton struggled from the start with the ratification process, and when George W. Bush took office in January 2001 things were at least clarified: in March 2001 he said that America had no intention of ratifying. This despite the publication of the third, and even more alarming, IPCC Scientific Assessment in 2001.[162]

Many people thought the Climate Convention was dead. But in a landmark decision for international relations, at the annual climate summit required by the convention, governments voted in November 2001 to go ahead without the US.

Many countries subsequently ratified the protocol. All eyes were on Russia. Time dragged by. For a long time, it looked as if President Putin wouldn't do it. But in October 2004 the Russian duma finally ratified the protocol, meaning that it came into force in February 2005.

One of the mechanisms negotiated into the protocol by governments involves the trading of emissions credits. This will involve both governments and companies. The idea is that if you are behind on achieving your agreed targets, either a national target in the case of governments, or an in-country target in the case of companies, you can buy emissions' credits from entities who are ahead of their targets. The EU, Japan and Canada are all on track to fall short of their targets, and are going to be in the market for emissions' trades. Russia, having shifted from a command to a market economy only since 1990, leads those in the market to sell.[163] The EU began its own emissions' trading scheme in January 2005. A global scheme is expected by 2008. Estimates vary for how valuable the carbon markets will prove to be, but they could be substantial. One estimate for the European market is 10 billion euros by 2007.[164] In any event, they seem sure to be a major new feature of corporate and intergovernmental energy policy in the coming years.

The contrast between Europe and America on this issue could hardly be starker. George W. Bush, re-elected in November 2004, maintains course with his unilateralism, in this as in other areas of international relations. Much of US industry, led by ExxonMobil, holds the same line. In the UK, by contrast, the Royal Commission on Environmental Pollution – a panel of experts drawn from government, industry and academia – urged in 2000 that the national government adopt the ambitious goal of 60 percent cuts in emissions as a policy target by 2050, so seriously did they view the global-warming threat. In a far-reaching consultation for the UK government's Energy White Paper, published in 2003, not a single company in the British energy sector questioned the Royal Commission's premise. The government duly adopted the target. Although assumed on behalf of a government decades from now, this target is testimony to the depth of concern in the UK. Prime Minister Tony Blair views global warming as the most important of global problems to solve, and has made the issue a priority for his chairmanship of the G8, and presidency of the EU, in the first six months of 2005.

That said, and to conclude, the international response to the global-warming problem has been on the whole pitiful compared to the obvious scale of the threat. If somebody had said to me in 1990, as I enthusiastically began campaigning for Greenpeace International on climate change, that the world would have achieved nothing of substance on emissions' reductions by 2005, and instead be staring down the barrel – as things stand – of steady and fast growth in global emissions, I would have dismissed the notion as impractically cynical. In my naivety, and surrounded as I was then by an ocean of rhetoric from world leaders and leading politicians about the enormity of the problem, I felt sure the sands would shift.

History is not destiny, as I have said before, and enough of a political critical mass could yet emerge for effective policy action on climate change. But today I am battle hardened. I am no longer sure that global warming is enough of a real and present danger to shift the fossil-fuel status quo on its own.

This is where the issue meets head-on with oil depletion. Society is going to be *forced* to deal with that problem, and soon.

Here, however, there is no rhetoric from world leaders and leading politicians, much less policy action as yet. Most of them have probably never even heard of the problem.

CONFLATING OIL DEPLETION AND GLOBAL WARMING: HOW MUCH MORE FOSSIL FUEL CAN WE "AFFORD" TO BURN?

Early toppers have had a tendency to ignore or downplay global warming. Launching a report in October 2003, for example, a team from the University of Uppsala told CNN that global warming would never involve a "doomsday scenario", because oil and gas are running out faster than most think.[165] This is wrong, sadly, as I show below.

Before explaining why, I should also add that environmentalists have had a tendency to downplay or ignore oil depletion, and still do. This may be due to a lack of the geological knowledge needed to appreciate the power of the arguments. But I have to hold my hand up here. I really did not appreciate the extent of the problem until the Shell reserves fiasco made me decide to look at it in more detail at the beginning of 2004. I have also heard the view from environmentalists that the issue is too depressing: if the early toppers are right, the world will leap to coal as an alternative and forget about global warming. That is a major issue, to which I will return.

The carbon arithmetic

How much carbon can we "afford" to burn, and how do the figures compare against early and late oil and gas topping points? We have to make a number of assumptions in attempting to answer this question. First let us assume that we want to stay below the 2°C threshold of global-warming danger identified by the European Union. In reality, the threshold may be lower. Then let us assume that the sensitivity of the climate to the heat-trapping abilities of the greenhouse gases is as it was calculated most recently by the Intergovernmental Panel on Climate Change. It could be higher, because of excess but unquantifiable emissions from amplifying feedbacks. If we make all these assumptions, we can "afford" to burn not more than 400 billion more tonnes of carbon from fossil fuels.[166] (The caveats force us to put the word "afford" in inverted commas. Whatever we do, we gamble to some degree. We burn carbon at something in excess of 7 billion tonnes a year from fossil fuels at present.)

How does this compare against the available "resource" of oil, gas and coal? Again, we are forced to revise our view of fossil fuels as "resources" in the strict sense. They could easily be a means of committing economic and environmental suicide, if overused. Let us assume for the moment, however, that the late toppers are right and that there are a trillion barrels in oil reserves, and another trillion left to find, in round numbers. This amounts to a total "resource" of some 270 billion tonnes of carbon. On top of this, the IPCC accounts 440 billion tonnes of "resource" in unconventional oil, making more than 700 in all. Then comes 500 plus billion tonnes of carbon in gas, excluding any methane hydrates. On top of the 1,200 plus billion tonnes of oil and gas comes coal, where the total exceeds 3,500 billion tonnes.[167]

The implication is clear: we cannot afford to burn all the oil, much of the gas must remain below ground, and the great majority of the coal shouldn't even be considered. Unless, that is, we want to see what business-as-usual build-up of carbon dioxide in our atmosphere will do to our civilization, and find out whether the comparisons to weapons of mass destruction apply.

Thinking of the challenge in terms of how much carbon has to stay below ground in fossil fuels is one approach. Another is calculating how much renewable energy capacity would have to be brought onstream to replace fossil fuels. Earlier we saw that 400 ppm of CO_2 in the atmosphere is a good candidate for the level to be exceeded at our absolute peril. A group of American researchers from the Union of Concerned Scientists and the US National Renewable Energy Laboratory have summarized what it means, in terms of renewable capacity, to stay below various levels of atmospheric CO_2 concentration. The results are daunting, to say the least. Even to stabilize concentrations at 750 ppm – far above the danger threshold – we would have to be installing 450 megawatts of new renewable capacity every day for decades to come. To hit a 550 ppm stabilization level, also well above danger, we would need 920 megawatts daily. Such a level of installation, maintained until 2050, would mean 49 percent of renewables in the global energy mix.[168] I emphasize, we would still be well within a voyage of exploration into how dangerous global warming is even if we hit these ambitious targets. I return to this awesome challenge in the last chapter of the book.

The spectre of a rush to coal

The lures of coal, at least for those who think nature is somehow a manageable entity, will be huge. First, liquids that can be burnt instead of oil can be extracted from coal. Second, much electricity is already derived from coal. Third, coal advocates suggest that emissions from coal-burning or liquefaction operations can be "sequestered" – buried below ground or stored in some other way without being vented to the atmosphere.

The coal-to-liquids technique derives from the Fischer–Tropsch process, a technique developed in oil-strapped Nazi Germany during the Second World War. In this process, steam and oxygen are passed over coal at high temperatures and pressures. Hydrogen and carbon monoxide are produced and then combined chemically into liquid fuels. China has recently signed an agreement with a South African company to build two liquefaction plants, at $3 billion each, producing a total of 440 million barrels a year. Cost has long been a barrier to developing such plants but, with total costs reportedly now at $15 per barrel, the attractions compared to current oil prices are obvious, if you are prepared to forget about environmental considerations.[169]

The prospect of many such plants proliferating around the world is real, because so many countries use so much coal already in electricity generation and reserves are so vast. The average EU coal share in power generation is 27 percent. In the US it is almost 50 percent, in China more than 76 percent, and in South Africa and Poland more than 90 percent.[170] The country with the most coal reserves is the US, which is also the country where pressures to find or make gasoline at almost any cost may prove to be the highest.

Advocates of underground storage for carbon dioxide from coal burning can be found in growing numbers in the highest places. The Chairman of Shell, Lord Oxburgh, recently gave an interview covering not just Shell's immediate woes on reserves, but global warming too. "I'm really very worried for the planet," he said, to the consternation of many in his industry. "You can't slip a piece of paper between David King [the government science adviser who thinks global warming is a worse threat than weapons of mass destruction] and me on this position." In the same interview he also said "… if we don't have sequestration then I see very little hope for the world."[171]

But what of renewable energy in that regard? Shell owns one of the largest renewable-energy companies in the world. From Lord Oxburgh's long interview, you would never have guessed it.

3 November 2004, Berlin

My mum, the royalist, is proud of me. The Queen is on a state visit to Germany. She has gone public and said she is very worried about global warming. She wants her experts to meet with their German counterparts in Berlin, while she is there. She wants them to tell her, her Prime Minister and the German Chancellor what they think is going on and what can be done about it. I have been invited.

I wend my way through tight security into the fortress-like British Embassy. Inside, coffee and croissants are being served as the two delegations await Her Majesty's arrival. I mill around, greeting old acquaintances. One is Sir John Houghton, former Chairman of the Intergovernmental Panel on Climate Change Scientific Working Group. Sir John, now retired, used to head the British Meteorological Office. We met in 1990, when he chaired the negotiations on the wording of the First IPCC Scientific Assessment of 1990, the report that kicked off the fourteen years of climate negotiations since. He had stayed in that role for the Second IPCC Assessment of 1995, which re-energized the negotiations with the urgency of its warning, and led ultimately to the Kyoto Protocol in 1997. Through it all he had had to marshal the considerable egos of several thousand of the world's best climate scientists, from dozens of countries and cultures, not to mention handle the horde of fossil-fuel lobbyists and their diplomatic proxies who were trying every dirty trick in the schools of disinformation and obfuscation to shoot him and his process down. I am talking to a man to whom the world owes a debt.

"I bet you're not really retired," I venture. He smiles, and tells me his latest project. Sir John, a Christian, is trying to convert the American Evangelicals to the view that we affront God if we trash the planet, and that by fuelling global warming we are doing just that. He describes a series of meetings he has had and how encouraged he is by the outcome. Many key Evangelical leaders have signed a covenant to do something about the problem, he says, eyes shining. This includes action in Congress. He is visibly excited, and so he should be. If American Evangelism is not a crucial constituency in turning things around I don't know what is.[172]

There is something about Sir John's manner now that I don't remember from the 1990s. A greater sense of urgency. A whiff of passion, even. During his long service with the IPCC he was a scientific diplomat: a reserved and cautious man. In some of our first interactions, I had urged him to emphasize the worst-case analysis of global warming: a coalescence of amplifying feedbacks that would make the eventual warming even more terrifying than the horrible situation estimated by the climate models then in use by all the national research centres contributing to the IPCC. Such feedbacks might even generate a runaway effect, and the IPCC should say as much, I argued. No, was his response. The IPCC's brief from the United Nations was to come up with a "best estimate", no more, no less. Anything else was just scaremongering. Now, fourteen years on, with scientific assessments significantly more sophisticated and appreciably more scary, it is clear what should have been done in 1990. I have been arguing for more than a decade that it was a major mistake not to give policymakers the worst-case analysis, as happens routinely in the world of military threat assessment. But there is no point in an "I told you so" conversation with Sir John, not today, not any other time. I hold my decade-plus of frustrations within.

Instead I tell Sir John about my recent trip to a global-warming conference in Tuscany. I know he will be interested, and he is. "It was like time travel," I say. Al Gore came to give a speech. Just like his speeches in the early 1990s, it was mostly about the climate science. He liked then to show off what he knew about the science, and little has changed, it seems. A man who has spent eight years as vice president of the United States you might think would have a few interesting things to say about the challenges of global-warming policymaking. But he rushed through "what to do about it" in a few minutes at the end of his talk. Just incredible.

I then tell Sir John that the most problematic of the contrarians he had had to deal with in the IPCC, Richard Lindzen, was at the conference. More time travel. Lindzen, a professor of atmospheric physics at MIT, was the only world-class scientist among the half-dozen or so who had hogged the news with their naysaying while Sir John was pulling together years of IPCC assessments. These sceptics all had their pet theories about why the thousands of scientists involved in Sir John's consensus reports were wrong, and how the proven heat-trapped character of the greenhouse gases pouring into the atmosphere from fossil-fuel burning would somehow be naturally counteracted. Lindzen's theory had to be taken far more seriously than those of the other contrarians, because of his status. It involved cloud formation in a manner that would wring heat out of the atmosphere in a massive negative feedback. Lindzen, despite being in a minority of one among the scientists active in the field, had made life difficult for Sir John, I knew. He had certainly been much used in the media by the fossil-fuel lobbyists of the Carbon Club. His performance in Tuscany had been just like those I had seen in the early 1990s, with a difference: this time he had compared his role in the climate-science debate to Galileo's lone struggle with the church. This I know will send Sir John ballistic, but there is more. Tilting at the IPCC, Lindzen had actually quoted Goebbels: "There is little doubt that repetition makes people more likely to believe things for which there is no basis."[173]

Sir John tells me that Lindzen had recently been wheeled out in Moscow on the side of the Russians trying to stop the ratification of the Kyoto Protocol. He is in full flow about the morals of hired-gun scientists as two gentlemen drift up to say hello to us. One is Lord Ron Oxburgh, Chairman of Shell and one of my old mentors at Oxford. Ron's career and mine have had similar trajectories, his inside the mainstream and mine outside. While I was Chief Scientist at Greenpeace UK, in my first year with them, he was Chief Scientist at the Ministry of Defence. While I have been a leader in the solar industry, he has been a leader in the oil industry. I like and respect Ron. He is aghast about global warming, and desperate to do something about it, whatever I might think of his policy proposals. The other man, who I don't know, Ron introduces as Kurt Dohmel, Chairman of Shell Germany.

"What is Shell doing about global warming?", Sir John asks as soon as the introductions are over. I'm right about the change in him, I decide. He now has the air of a man driven by passion born of impatience. He has become radicalized by his decade of having to deal with foot-dragging Americans, oil-industry lobbyists, OPEC diplomats posing as scientists and the like. The Sir John Houghton of old would have been far more circumspect.

The two Shell men assure Sir John that renewable energy is becoming ever more important in their company. Dohmel tells him that they have a big solar photovoltaic manufacturing plant in Gelsen-kirchen. It is sold out a year ahead. Demand is way ahead of supply.

"So why don't you build a bigger one?" Sir John says, almost rudely.

I imagine Dohmel is about to tell him that they are in the process of doing so. Certainly every other PV manufacturer in the world is scaling up fast. But no.

"Because it takes four years and then they are out date."

Sir John Houghton looks at me with an exasperation he does not bother to hide.

CHAPTER 6

How we got into this mess, part 1: Before the knowledge

Before we can work out the best route to limiting the damage and escaping from the traps we have built for ourselves, we need to understand the history of how we got here. I have tried to make a distinction in the narration of that history over this chapter and the next between the years before unignorable knowledge (in other words before the early 1990s) and our actions in what I shall call the years of complicity. But it is important throughout not to lose sight of the big overarching story. The main oil companies are now bigger and more powerful than most governments. They have developed intimate relationships with the governments at the centre of the crisis. These institutions are not going to evaporate and leave ordinary citizens to solve the problems. They have to change, themselves, in the process. They have to become more part of the solution than of the problem. The cultures that have evolved within corporations and governments will have a lot to do with the extent to which this can happen, and how fast.

In the case of the big oil companies, their cultures have evolved over more than a century. Exxon, BP, Shell, Mobil, Chevron, Texaco and Gulf were all massive global oil corporations before the First World War broke out. They controlled everything they needed in order to be stand-alone in the oil business around the world, from oilfields to tankers, to refineries and distribution outlets. They worked so closely together to build up their transnational empires, all that time ago, that they were effectively one empire, and became known as the "seven sisters". They weren't labelled with their current brands at the time, but I will mostly call the originating companies by their modern names in what follows.

UN-AMERICAN ACTIVITIES: 1859–1911

There were laws then, as there are now, against companies becoming so big and powerful that they can mistreat the small guy with impunity. They are called anti-trust laws. Exxon's predecessor, Standard Oil, fell foul of them in 1911. The company operated a billion-dollar international cartel that exercised a truly cut-throat attitude to any small players seeking to trade in oil. In the process they bribed senators and congressmen and ran a network of spies around the world. In the first decade of the twentieth century, such a portfolio of activities was considered, by millions of Americans, to be un-American. A journalist called Ida Tarbell, whose father had been ruined by Exxon while trying to produce oil where it was first found, in Pennsylvania, wrote a book about the company's antics that became a bestseller. Ida's book has been called the most influential book on business ever published.[174] Largely because of this book, Exxon was viewed as not having the national interest at heart: of being an uncaring giant that trampled on the American entrepreneur using tactics that were certainly unethical, and frequently illegal.

Exxon's then boss, John Archbold, tried to bribe the then American President, Theodore Roosevelt, with a campaign contribution equivalent to $2 million of modern money.[175] Yes, they called it bribery in those days. Roosevelt sent it back, mighty affronted, and told his Attorney General to get the hell on with applying the anti-trust laws. In 2000, ExxonMobil sent George W. Bush a campaign contribution of a million dollars plus.[176] He said thanks very much.

Faced with Roosevelt's wrath, a senior Standard Oil executive wrote an advisory memo as follows: "I think this anti-trust fever is a craze which we should meet in a very dignified way with answers which while perfectly truthful are evasive of bottom facts." An interesting definition of truth here. Truth with omissions. Bear this in mind when I get to ExxonMobil's performance on global warming below.

Whatever, machinations were no use. ExxonMobil's predecessor was indicted. They appealed. The Supreme Court upheld the conviction. The company was ordered to divest itself of all its subsidiaries, in other words to split itself up into small companies, within six months. They were even banned from drilling in Texas.

It was a big deal at the time, but it wound up making little difference. The oil industry was growing too fast. Henry Ford had worked out how to mass-produce the horseless carriage. Roads were being built everywhere. The navies had switched from coal to oil, and war was looming. Chopping up Standard Oil was like trying to lop heads off a hydra. Standard Oil's holding company would later become Exxon. Standard Oil of New York became Socony (later Mobil). Standard Oil of California became Socal (later Chevron). Standard Oil of Indiana became Amoco. Were they separate companies? In 1915, the Federal Trade Commission found that the "Standard Group" operated a common understanding, and was still able to curb competition.

How many giant oil companies are there now in the world? After an orgy of mergers in the 1990s, we have five: ExxonMobil, BP (into which

Amoco and others have been subsumed), ChevronTexaco, TotalFinaElf and Shell, who somehow didn't get around to gobbling anyone significant up. So much for nearly a century's worth of anti-trust laws. We are right back where we started a hundred years ago, except that now most Americans would probably not consider ExxonMobil at all un-American. The difference is that at the beginning of the twenty-first century the corporations are far more powerful than they were at the dawn of the twentieth. If you combine profits measured in millions of dollars an hour with a hundred years of vested-interest institution building, what else would you expect? At the beginning of the twentieth century, at least in the United States, people worried about the government and the power it had to rein the oil companies in. Now, in the number one oil-consuming nation, the oil companies essentially *are* the government.[177]

CONSPIRACY AND TREASON: 1928–1945

It's a shame that nobody ever again had the will Theodore Roosevelt and his Attorney General did, because the anti-trust laws became much needed as time went by. Until 1928, the "big three" – Exxon, Shell and BP – had competed just like they were supposed to as good capitalist flagships. As a result, they had driven the world to a glut of oil and low prices. However, low prices are not good for profits. Their leaders decided to do something about that. They met in a castle in Scotland and cooked up an agreement, the outrageous details of which the world would not learn about for more than thirty years. They agreed to put an end to price-cutting and over-production. They decided all three would accept the volume of their present share of the international oil business going forward, and that they would share any future increase in production in like proportion. They called it the "As Is" agreement. They even agreed a pricing formula.

To say that this transgressed the US anti-trust legislation that had brought down Standard Oil – temporarily – is an understatement. No wonder they kept it so secret. It was an affront to democracy: an award by a handful of moguls, to themselves, of the right to allocate the oil trade, to fix prices, and to do so globally. At least Enron only rigged the *California* energy market in 2000. In 1928, this lot rigged the world. One of them – Henri Deterding, President of Shell – was later to join the Nazi party. When he died six months before the war started, Hitler and Goering sent wreaths for his funeral. Another – Walter Teagle, the President of Exxon – exchanged information with a German chemical company after Hitler came to power, allowing the Nazis access to patents vital for the manufacture of aviation fuel. Teagle continued to trade vital information even after Hitler had invaded Poland. He blocked US synthetic rubber research, an area in which the company held patents, around the time that the Japanese were over-running the Malayan rubber plantations. For this, he stood accused by US President Harry Truman of treason. Teagle said of Communists that they repudiated the civilized code of ethics. He was a man of interesting ethics himself.

Far from being anathema to the other oil companies, the "As Is" agreement was later adopted by fifteen other American companies, including the other four US sisters. The companies had themselves a secret global cartel.

The US Department of Justice, agitated about the treasonous attempt to profiteer in rubber, had another crack at Exxon with the anti-trust laws in 1940, but they had picked a rather bad time. The shooting was about to start again, for the Americans. The government was going to need the oil companies. You couldn't fight a war without lots of oil, as the Germans and Japanese were about to discover.

By this stage, the oil industry had a robust and faithful bedfellow, a business growing just as fast as oil: the auto industry. The auto companies were also not averse to a bit of conspiracy in advancing their markets. In the late 1930s, General Motors, along with Chevron, bought up the electric railroad system around which the suburbs of Los Angeles had been designed. They then closed it down. They did the same thing, with other oil companies, in other big American cities. People and their local governments were being locked into an arranged marriage with oil and autos whether they wanted it or not.

SEND A GUNBOAT: 1951–1956

The forerunner of BP, the Anglo Persian Oil Company, had discovered huge oil deposits in Iran, or Persia as it then was, as long ago as 1908. But the then Shah of Persia was a pretty unlovable, and hence unloved, figure. In 1951, after a round of strikes, martial law, overtime by the secret police and all the sort of things that go on in the run-up to tyrants falling, Iran actually held elections. A populist leader, a Mr Mossadegh, was voted in as Prime Minister. He moved immediately to seize BP's oilfields.

The British Labour government, despite having given Iran clear models for nationalization in its own domestic policy, now instantly took with Colonel Blimpish ire to its gunboats. The Foreign Secretary, Defence Minister and various senior armed services advisers all advocated military action to seize the oilfields back. The British navy was dispatched to patrol offshore Iran and blockade oil exports. A parachute brigade was readied for action. But, in the White House, President Truman signalled a big red stop light across the Atlantic. The British backed off. They missed out on that particular military misadventure, but made up for it spectacularly a few years later in 1956, when they invaded Egypt to seize the Suez Canal.

So Saddam Hussein's invasions of Iran in 1980 and Kuwait in 1990 were not the first invasions of a Middle Eastern country over oil. One has to wonder how many enthusiastic British supporters of the invasion of Iraq in 2003 knew that a British government of the past had threatened to invade a Middle Eastern country which had nationalized its own oil, and then actually did invade a Middle Eastern country which had nationalized a canal carrying oil tankers through its own territory?

What happened to Mr Mossadegh? Not good things. The CIA got rid of him in a coup that cost them less than a million dollars to engineer.[178] Far less expensive and a lot less messy than a war. Why couldn't they have tried that one on Saddam Hussein?

THE OIL PRODUCERS ORGANIZE A CARTEL TO CONFRONT THE CARTEL: 1960–1972

Huge amounts of oil were found in the Middle Eastern oil states during the 1930s, '40s and '50s. The oil companies, operating their cosy secret cartel, forked out enough money for the discoveries to turn poor tribal chiefs into some of the richest people in the world. But they still paid only a pittance, compared to what they made themselves when selling on to the now thoroughly addicted populations of America and Europe. The producers knew this, and in 1960 began trying to do something about it. Saudi Arabia, Iran, Iraq, Kuwait and Venezuela formed the Organization of the Petroleum Exporting Countries, otherwise known as OPEC. Other oil producers joined later.

Their goal was to hike the price through co-ordinated action, but they were a quarrelsome alliance, and it took them a long time to get organized. They weren't helped by the fact that America had not yet reached its own peak of oil production, and didn't need imports anything like as badly as it would later, in the 1970s and beyond.

In 1967 Israel invaded Egypt. The Arab states became predictably furious. They decided to use what they called the "oil weapon": a co-ordinated turning down, or even off, of the oil taps. But before this embargo had a chance of taking effect, Iran and Venezuela broke it. A former Exxon executive summed up the attitude of the day. Exxon, he said, "seemed certain that the Arabs would never get together. Their image of the Arabs was taken from the film *Lawrence of Arabia*".

Well, Exxon was wrong.

THE FIRST OIL SHOCK: 1973

All through the account that follows, it is worth thinking about what history would have been like if, instead of the somewhat Westernized royals in the House of Saud, the West had been dealing with a funda-mentalist regime made up of Saudis who hated America and Britain as much as the average Saudi citizen does today.

In 1971 the OPEC countries managed finally to sing from the same sheet. Meanwhile, US domestic oil production had just peaked and begun to decline. At a meeting in Tehran, OPEC oil ministers agreed to a hefty hike in the price of oil. There was little the companies or their governments could do about it. The ministers, led by Saudi Arabia's suave and savvy Sheik Yamani, also talked about trying again with the oil weapon, should the need arise. Yamani was a master when it came to balancing stick and carrot. His goal was "participation" with the oil companies: an "indissoluble marriage", as he put it.

The Shah of Iran joined in enthusiastically. He told the *New York Times* in 1973 that a significant increase in the price of oil was inevitable, and justifiable. "You increased the price of wheat you sell us by 300 percent, and the same for sugar and cement ... You buy our crude oil and sell it back to us, redefined as petrochemicals, at a hundred times the price you've paid to us ... It's only fair that, from now on, you should pay more for oil. Let's say ten times more."[179]

There have been six oil price peaks since 1880, all of them since 1973. Figure 1 on p. 26 shows them, with the price of oil in dollars of the day and in 2003 dollars. We need to understand all these peaks and what happened after them so as to understand the impacts of the current burgeoning oil-price crisis. The peaks are numbered in Figure 1 so we can consider each in turn.

In 1973, Israel and her neighbours set about each other in the Yom Kippur War. This time Egypt and Syria went first, trying to grab back the land Israel had seized in 1967. The Egyptian leader, Anwar Sadat, was a smart operator, and had teed up the Saudi leader, King Faisal, to back his armies up with the oil weapon. The then American President, Richard Nixon, was preoccupied at the time about the brewing Watergate scandal – the discovery that he had sanctioned a break-in by amateur burglars of the Democrats' headquarters building, aiming to plant bugs – that would eventually cause his impeachment, and so did nothing to appease the Saudi King. Faisal then proceeded – along with the rest of the Middle East oil states – to embargo all supplies to the US and the Netherlands, and to instigate cutbacks in supply for all countries unfriendly to the Arab cause.

The Saudis wielded their oil weapon for only a few months. At its worst point the embargo led to a net drop in world oil supply of only 4.4 million barrels per day in a market then consuming around 50, some 9 percent of the total.[180] But even this tiny amount was enough to cause havoc. Though the oil price was lower than it is at the time of writing in today's dollars, it sent shock waves around the world. Fear of imminent economic collapse leapt out of nowhere. Consumer stockpiling became rampant, compounding the problems. Americans and Europeans had to queue for gasoline for the first time in history. All driving was cancelled on Sundays in Germany, where ramblers appeared on deserted motorways. In the UK, where the oil-price hike coincided with industrial action by coal miners, the government had to impose a three-day working week because there wasn't enough energy supply around to allow the country to work a normal five-day week. In the US, a national speed limit of 55 miles per hour was imposed to help reduce consumption, and consumers abandoned Buicks in favour of smaller cars imported from Japan.[181] Power cuts became a fact of life.

Oil had become a weapon indeed. The embargo ended after just a few months in major part because Saudi fury over America's support for Israel became tempered by a fear that their embargo would create a global depression, so hurting themselves as well as everyone else. As Sheik Yamani told the West: "… if you go down, we would go down."[182] A ghastly recession followed the 1973 embargo, but the world economy did not tip into depression.

The Middle East-related geopolitics of the time were a crazy inverse of the modern era. As the gasoline queues built up in the States, the US threatened to invade Arabia. Beleaguered governments often seek wars to get them off hooks, and the Nixon government knew it was playing to a well-primed audience in this particular episode of sabre rattling under domestic duress. As a US Defense Department official put it: "… the Naval War College was filled with Marine Corps colonels walking round saying: 'We're going to put those Goddamned rag heads back on their camels.'" British government documents later declassified show they got as far as counting the troops it would take to do the job – two brigades to tackle the Saudis and one each for Kuwait and Abu Dhabi.[183] As US defence secretary James Schlesinger later put it: "Militarily we could have seized one of the Arab states. And the plan did indeed scare and anger them. No, it wasn't just bravado. It was clearly intended as a warning. I think the Arabs were quite worried about it after '73."[184]

Evidently they were worried about it during 1973 as well. The Saudi King instructed his National Guard to rig their critical oil installations to blow in the event of a US invasion. He was evidently prepared to cut off his nose to spite his face. The source for this revelation says it would have taken the US a year to bring the country back up to full production.[185] What a contrast this creates to the events associated with the first Gulf War in 1991, when Saddam Hussein – an invader, not an incumbent regime – blew up the Kuwaiti oilfields as George Bush Senior's America advanced.

What would America try to do today if the House of Saud fell to people more angry with Uncle Sam than King Faisal was in 1973? People who might perhaps prefer to sell their oil to the Chinese and Indians, with their soaring demand, and to hell with the Great Satan? How would the conversations be going in the Naval War College and the Republican Party today, if that happened?

We don't need to recall the events of 1973 to know the answer to that one. And it isn't pretty.

BACKLASH AFTER THE FIRST SHOCK: BIG OIL WOBBLES

The thing about high oil prices, as things stood then and stand now, is that they mean economic pain for almost everyone except the oil companies. As the oil shortage grew worse in the cold winter of 1973–1974, Big Oil announced record profits. Exxon's annual total was an all-time global record for any corporation: $2.5 billion. This went down with consumers like a lead balloon. The companies launched aggressive advertising campaigns to justify their huge profits, saying they needed to recycle them into costly new exploration. But Mobil ruined it all by recycling half a billion of their profits out of oil, and indeed out of energy: they bought a chain of stores.

Opinion polls showed that Americans did not believe the companies. They blamed them more than the Arabs. A hilarious and popular lapel badge appeared on the streets: "Impeach Nixxon".

In March 1974 the embargo ended, and not long thereafter Nixxon duly resigned. The inquests began in earnest. The most colourful assault on the companies came in Senate hearings held by Senator "Scoop" Jackson's Permanent Subcommittee on Investigations. Representatives of the "seven sisters" gave dismal televised performances when questioned. "The American people want to know if this so-called energy crisis is only a pretext," Jackson fumed, "a cover to eliminate the major source of price competition – the independents; to raise prices; to repeal environmental laws." His persistent reference to "obscene profits" gave Americans a new national catchphrase.

Incredibly, 45 out of 100 senators on Capitol Hill voted for divestiture of the oil companies. Divestiture would have involved the cutting of each of the integrated companies into totally separate companies for production, transportation, refining and marketing. The oilmen called it "dismemberment", and that is what it would have been.

By the time the Arab oil embargo was lifted, American auto manufacturers were beginning to retool to make smaller cars. City planners were beginning to look hard at mass-transit schemes. The price of suburban real estate was falling through the floor. All these were developments that could have moved the US, then, away from "dependence" on Middle East oil. They could also have made more than a dent in the subsequent flow of greenhouse gases into the atmosphere.

The companies flirted with a serious change of course in the aftermath of 1973. Oil was getting more difficult to find and extract, even then. In 1975, the first hard-won oil came ashore from the stormy North Sea, but it had taken six years since discovery. In 1976 a planning document was put before the board of one of the largest oil companies warning that "the future for private oil companies is much less certain. The trend for upstream operations to pass into government hands will continue ... Greater government involvement, direct or indirect, is also to be expected downstream."[186] The companies continued to spend heavily on exploration. But they also, for the first time, began to move huge sums into completely separate fields. Coal, chemicals and copper were among the commodities in which they invested. Shell and Gulf even invested heavily in nuclear power. Almost all this diversification failed in due course. Oil men have often repeated a mantra to me from those days. It goes something like this: "We are in the oil business, that is what we are good at. We should stick to our core products, and our core business." And since the 1980s, that is exactly what they have done.

But history shows they can be driven to the edge of enforced change.

THE INDISSOLUBLE MARRIAGE: 1974–1978

In the first three months of the embargo, US gasoline consumption had decreased by more than 7 percent. But, after the embargo ended, consumption shot right back up again. Apathy instantly reasserted itself in Congress. Almost eight hundred potential bills concerning energy had been circulating on Capitol Hill during the oil crisis, many of them intent on promoting the prospects for renewable energy, renewable fuels and energy efficiency. But by July 1974 only eight of them had become law, and two of those involved the suspension of pollution laws and the opening of the Trans-Alaska oil pipeline. President Ford, a man with an apposite name, was in the White House. He now almost halved the budget on a bill designed to promote mass-transit in cities. Soon afterwards, playing to the auto lobby, he said, "I am a Michigander, and my name is Ford."

In April of 1975, the US Securities and Exchange Commission, investigating political "contributions" by the "sisters", forced Gulf to admit that between 1960 and 1973 over $10 million of corporate funds had been offered as "contributions" in the United States and abroad, often illegally. Exxon had to admit to having made "political contributions" overseas. These bribes had remained hidden for years, despite the scrutiny of auditors, thousands of shareholders and the Washington legislators – amid the huge sums of money in which the companies dealt – in their labyrinthine and opaque accounts.

In May of 1975, Exxon overtook General Motors to become the world's biggest corporation. The five US "sisters" shared the top seven spots in the Fortune list with General Motors and Ford. The basic truth was now clear. Governments had been quick to suspend their anti-trust apparatus during times of crisis like the carve-up of Iran, or in the run-up to the embargo, and now found themselves no longer effectively able to reapply it. Nobody knew how to take on a cartel which represented the "indissoluble marriage", to use Sheik Yamani's expression, of trans-national corporations and foreign producer governments.

THE SECOND OIL SHOCK: 1979–1981

The Shah of Persia, impatient to pursue a catalogue of crazy dreams in an increasingly disaffected country, began professing that oil would not be overpriced if it sold at $100 a barrel. As he was fond of arguing, Coca-Cola cost $89 a barrel. The Saudis took a different view. They did not want to overcook their pricing of oil. Yamani would often say how little it would take to make the West launch a serious drive towards alternatives. As he volunteered in 1981, "… if we force Western countries to invest heavily in finding alternative sources of energy, they will. This will take them no more than seven to ten years and will result in their reduced dependence on oil as a source of energy to a point which will jeopardize Saudi Arabia's interests."[187]

Seven to ten years? If we could have done it then, we can certainly do it now.

Revolution, meanwhile, was afoot in Iran. Despite a secret police force even more ferocious than his father's, and armed forces tooled up by the best America and Britain could offer, the Shah fell in early 1979.[188] A popular fundamentalist Islamic regime replaced him. Royal families in Saudi Arabia and Kuwait pray to this day that this has not set a precedent for the replacement of tyrant monarchs by fundamentalists. Osama bin-Ladin and his followers have ensured that their prayers are not guaranteed to produce the desired result.

Iranian oil production inevitably fell during the revolution, and up the price went again. At more than $80 a barrel in today's money (peak 2 in Figure 1 on p. 26), the resulting 1979 oil peak was the worst there has ever been. It also lasted longer than the first. The crisis was compounded when Iraq's new leader Saddam Hussein invaded Iran in an audacious oil grab in 1980, aiming to exploit what he thought would be internal chaos as the Ayatollah Khomeini took over the Iranian government. He got it wrong. The Iranians fought back with gusto, threw human waves of fanatical children at Saddam's guns, and he ended up losing much of his own oil production. The disrupting impact that had on supplies in both nations kept prices high until late in 1981.

This crisis, like the first, involved only a small amount of oil. It seems incredible with the benefit of hindsight that there was so much economic turmoil at the time when you consider that net production in the non-

Communist world fell by only 2 million barrels per day as the revolution played out in the first quarter of 1979, some 4 percent of the global total. So why did the price go from $13 to fully $34 so quickly then? A major part of the answer involved a rather human ingredient: pure panic.

This is a point I need to emphasize. This is what we tend to do, collectively, when oil supplies dry up just a little. We panic.

Nobody knew whether the fundamentalist coup in Iran would spread to other Middle Eastern producers, or how the incoming zealots would play their oil card. So everyone stockpiled: companies, heavy commercial users, trading companies and consumers alike. The world quickly locked itself into a complex vicious circle. American gas lines went round the block again, the idling cars wasting 150,000 barrels a day standing still, according to one estimate.[189]

This time the resulting global recession was accompanied by inflation. Voters, not understanding the complexities, blamed both the oil majors and governments. The 1979 crisis was the beginning of the end for the then US President, Jimmy Carter. He called the new imperative to save energy "the moral equivalent of war". He instigated a national synthetic fuels programme, told the American automobile industry to increase its average fuel efficiency, demanded energy-efficiency standards be set for buildings and put solar panels on the White House. None of it did him any good. He lost the 1980 election to Ronald Reagan, a firm friend of the oil industry, who took down the solar panels on the White House.

The crisis ended late in 1981 for the reasons we discussed in Chapter 1. First, the Saudis – once again fearing a global economic depression – began pumping hard to compensate for the drop in Iranian and Iraqi oil supplies. Second, the stockpilers eventually started releasing inventories into the market, creating a glut. Third, oil had begun to flow from the huge discoveries in Alaska and the North Sea – decidedly non-Middle Eastern and hence much more secure locations.

As I have mentioned earlier, in the coming energy crisis none of these solutions pertain: the Saudis cannot pump faster like they did in 1980, there is no spare capacity elsewhere, little oil is stockpiled, and there is little likelihood of new discoveries on the scale able to meet fast-growing global demand. Bad as it was then, it will be much worse when the next crunch hits.

4 February 1980, Oxford, UK

I drag myself out of bed into the chill air. The litter of last night's candles lies around the bed. The power was down again, and still is. Little Jess is already awake, wrapped in a parka playing with her dolls, singing to herself. I turn on the Today *programme as I make tea. Tension-filled voices once again, trying to be calm but failing to mask the seething anger beneath the surface. Why can't you keep the lights on and the country running, Minister? Because the price of oil is sky high. The coal strike is biting. It's the greed of OPEC and the unions. Not our fault.*

Why are you striking just at the time the country is on its knees, Mr Union Official? Because they want to close the pits, take away our jobs, and replace us with gas men, Mr Smart-arse Interviewer.

I run through the timetable for the morning. Cycle to school, with Jess in the carrier seat. Drop her off, catch the train to London, hoping as ever that it will be on time so I can make it in time for my first lecture. This has been a big problem of late, but then so it has for all the commuters on the faculty at the Royal School of Mines. Nobody can complain too much. Besides, there is so much consulting going on. The place couldn't stand an audit of teaching conscientiousness. The oil companies are exploring everywhere. The geophysicists are helping them refine seismics. The geochemists are all working on source rocks. Even the palaeontologists are grafting for Big Oil, working on the dating of stratigraphic sequences in far-flung lands. I feel sorry for the students. They must feel like second-class citizens.

As I watch the toast I blow on my freezing hands and the breath comes out as steam. Then, cerrrrash! An explosion of sound from next door, a mix of ripping, banging and squelching. "Dad!" A seriously alarmed cry. I rush in. Little Jess is standing there in her parka looking at the ceiling, which is mostly on the floor around her blue ankles. Water drips steadily on to the pile, already turning it into mush. The water pipes have burst in the roof. With the heating down, they have frozen and cracked.

There will be neither lectures nor consulting for me at Imperial College today.

THE EXPLOITATION YEARS: 1981–1990

Goldman Sachs analyses the history of the oil industry since the 1960s as a cycle of "exploitation" and "investment" phases, marked respectively by low and high oil prices. Throughout the 1960s, at a time of low oil prices, the industry fed on the huge discoveries of the earlier decades without investing significantly in "next-generation" infrastructure. In 1973 an investment phase lasting ten years began, triggered by the high prices of the first oil shock and extended by the second. It was in these years that most of the infrastructure used today by the oil industry was put in place. It ended in 1983. After a transition period lasting four years from 1983 to 1986, a new exploitation phase lasting fifteen years began, with investment levels well below those of the 1970s.[190] As we have seen, Goldman Sachs's view is that the seeds of a major coming infrastructure crunch were being laid, unwittingly, in this period, meaning trouble even without an assumed early oil topping point looming.

So too were the seeds of the global-warming story. These were to emerge into full public bloom in 1990, with the advent of the IPCC's First Scientific Assessment Report, and the onset of the international negotiations for a treaty aiming to avert dangerous climate change. With the knowledge available from that point onwards, nothing about the use of oil could be viewed in the same way as before. I think of the years since 1990 as the complicity years, because this is the time during which we have had no excuses, collectively and individually, for continuing fossil-fuel profligacy. We have wilfully maintained course with our oil addiction, despite a very clear early warning that we are slowly destroying our own ability to live on the planet.

CHAPTER 7

How we got into this mess, part 2: The complicity years

GLOBAL WARMING: THE FIRST WARNING, AND THE EMERGENCE
OF THE CARBON CLUB

Anyone who reads a newspaper, much less the summary of a scientific report or two, must know that global warming is an issue with big implications. The balance-of-probabilities view has been clear since the completion of the first IPCC report, in 1990. On the day it was concluded, every UK newspaper carried front-page headlines about its conclusions. "Race to save the world", shouted one of the papers. The others may have been more sober, but they carried the same message.

From this starting point, living through the unfolding of relevant history on the cutting edge has been one very big learning experience for me. I was thirty-six years old and very well travelled in 1990, and yet – I am now forced to conclude – very naive. I had recently resigned from my lucrative oil consultancy and teaching at the Royal School of Mines on grounds of conscience. I couldn't face teaching aspiring young oilmen and women the tricks of their trade any longer. I thought I would be the first of many defectors from the oil and coal industries and their support cultures. I simply had no idea how deep and wide the river of denial would run.

A good deal of the complicity is indirect. There is much confusion about the greenhouse issue, and it is to a degree – with the benefit of hindsight – understandable. One of the reasons the confusion exists is the disinformation campaign waged from day one by the oil, coal, auto and related industries, and their supporters in governments. I was in the room where the world's top climate scientists completed the report that kick-started the climate negotiations, one of only two suitably qualified scientists from environment groups to be allowed in. I watched Exxon's man try to neuter the document that day aghast. Little did I know that this was only the beginning of a long catalogue of distortion, mis-information and lying by the oil industry and its allies over the decade to come. In 1998, I compiled the only eye-witness account of the history of the climate negotiations that has ever been written, to my knowledge. *The Carbon War* had a militaristic title because war was the way it felt to me, as a newly recruited environmentalist. I called the network of coal, oil and other fossil-fuel-related industry umbrella groups at the climate talks a milder name, the Carbon Club, but they fought dirty.

The groups all had anodyne names that disguised their purpose. Two organizations, the World Climate Council and the Global Climate Coalition, played particularly central roles. In August 1990, in Sundsvaal, Sweden, I watched Don Pearlman, leader of the World Climate Council, coach the Saudi Arabian delegation to the final IPCC meeting before the World Climate Conference. He sat in a hotel foyer, surrounded by senior Saudi diplomats, all with copies of the draft negotiating text of the IPCC summary report open in front of them, rehearsing tactics for watering it down, all within earshot of passers-by. The Saudis went from that seminar into the negotiations, where their stalling and filibustering included a concerted and barely credible effort to excise the words "carbon dioxide" from the document.

If this is the level that a government can descend to in defence of short-term interests – when longer-term stakes include a viable future for the planet – what hope can we have that they are telling the truth about depletion of their oil reserves?

THE FIRST GULF WAR (1990)

Throughout the 1980s, while the world's climate scientists were crystallizing their concerns about the effects of fossil-fuel burning on climate, oil from the North Sea and Alaska's North Slope flowed in abundance, and the oil price remained low for years. It next soared when Saddam Hussein invaded Kuwait on his second great oil grab, in 1990. You would have thought he might have learnt his lesson from the first one. The Iran–Iraq War ground on for eight years. But the Iraqi dictator must have figured he knew America quite well as a result of having regularly exchanged his oil for their weaponry, including chemicals, with which to fight the Iranians and suppress his own people.[191] He somehow came or was led to the view that America would not mind too much if he seized Kuwait.[192] He was wrong. The US and its allies counter-invaded Kuwait.

George Bush Senior said of America's first Gulf War: "It has every-thing to do with what religion embodies – good versus evil, right versus wrong, human dignity and freedom versus tyranny and oppression." How did he square that with the US threat in 1973 to invade Saudi Arabia for oil, I wonder?

The price peak caused by the Iraqi invasion of Kuwait was small and short lived (peak 3 in Figure 1, p. 26), but it was followed by economic recession, nonetheless, just like the earlier two.

THE OIL AND COAL INDUSTRIES IN THE RUN UP TO KYOTO (1991–1997)

The climate negotiations began in February 1991 with the first Gulf War in full swing. The first of the two-week sessions of negotiations were supposed to be in Washington DC, but Washington had more important things on its mind, and the talks were relegated to the backwoods of Virginia. Whereas a vast media circus had attended the World Climate Conference in November 1990, only a trickle of journalists bothered with the first session of proper negotiations, notwithstanding the presence of over a hundred governments. Unsurprisingly, little or no progress was made.

I have summarized in Chapter 5 the painfully slow path from the IPCC's first warning in 1990 to the Kyoto Protocol in 1997. Here let me offer some examples of how the oil industry, with its confederates, approached these potentially life-or-death negotiations.

In February 1992, at the UN in New York, the Global Climate Coalition (GCC) – a lobby group representing Exxon, Mobil, Texaco, Shell, BP and many other oil, coal and auto companies – used professional sceptic Fred Singer to attack the IPCC science at a press conference during the fifth session of negotiations. The GCC issued a briefing entitled "Stabilising carbon-dioxide emissions would have little environmental benefit". This kind of tactic became a feature of the negotiations: the deployment of one or more of a tiny group of well-known contrarians against the hundreds of scientists – most of them in government service, including the US's – who populated the IPCC side. It worked well. The media would almost always present the issue as a "balanced" debate, with one voice representing each side. Often, the single contrarian would be a media-friendly smooth talker, and the representative of the great majority an un-media-trained stumbler.

In June 1992, at the Earth Summit in Rio de Janeiro, with heads of state signing the newly negotiated Framework Convention on Climate Change by the dozen, the GCC staged a press conference which laid out a key Carbon Club wrecking strategy for the years to come. Executive Director John Schlaes led a concerted effort to emphasize the growing emissions of developing countries, preparing the way for a sustained attack by the Carbon Club on the soft underbelly of the Convention, which led to the many prevarications and potential deadlock between governments in the run-up to Kyoto. Yet the developed countries had conceded at the outset that most of the greenhouse gases in the atmosphere had come from emissions by the minority who lived in the *developed* world, and that it was they therefore who should step to the plate first when it came to legally binding emissions reductions targets.

In May 1993, I saw Bush Administration climate negotiator Harlan Watson lavish praise on the work of the GCC at an international coal-industry conference in Barcelona, warning the industry that it was gravely threatened because of the Convention on Climate Change, and exhorting delegates to work with the Carbon Club to defend their business. In 2004, at the time of writing, Harlan Watson heads George W. Bush's delegation to the continuing post-Kyoto talks.

In August 1993, at the eighth session of talks in Geneva, with the Clinton Administration's CO_2 policy still not clear, a World Coal Institute document headed "Fighting carbon dioxide would lead to economic disaster" plumbed new depths in disinformation, echoing a concurrent GCC theme – hotly contested – that emissions' limitations would cause massive US job losses.

In September 1993, on the same day that the President of the Reinsurance Association of America told an international insurance conference near Wall Street that global warming could bankrupt the insurance industry, Fred Singer wrote in the *New York Times* that "... as observations and theory diverge more strongly with each passing year, it becomes more certain there is something very wrong with the computer models that have been used to scare the world public and their governments into considering drastic, hasty actions." This exact reverse of reality was part of a shift of gears by the Carbon Club in its attack on the climate science, aiming to head off the potential for progressive moves by the Clinton Administration. "Conservatives and industry groups", the *New York Times* wrote, "have mounted a renewed assault on the idea that global warming is a serious and possibly catastrophic threat. In a drum roll of criticism over the last few months they have characterized the thesis of global warming as 'a flash in the pan,' 'hysteria,' 'scare talk,' and 'a ploy by socialists to justify controls on the economy.'"

In August 1994, at the tenth session of negotiations in Geneva, the largest-ever Carbon Club presence carried out a twin-pronged attack on science and competitiveness. "To date," John Schlaes wrote in a GCC submission, "science has been unable to establish what qualifies as a dangerous level of greenhouse gas concentrations. This makes a judgement on the adequacy of commitments logically impossible." Here was another recurrent theme: there is uncertainty therefore we need do nothing about emissions. The equivalent argument in the case of military threat assessment would have gone as follows: there was uncertainty about the Red Army's intent to invade Europe during the Cold War, therefore we had no need to spend money buying insurance via defence expenditure.

In February 1995, at the eleventh session of negotiations in New York, the GCC released a study by a weather consultancy, Accu-Weather, which claimed that there was no convincing observational evidence that extremes of temperature and rainfall were on the rise. We have seen how fallacious this was in Chapter 5. The temperature claim was based on three supposedly "representative" stations, all in the US, and the precipitation claim – incredibly – was based on just one station. As though one, or three, stations could represent the whole US. As though the US could represent the whole world. This was the single worst example of distortion I encountered. Imagine the same standards of science applied by oil companies in the oilfields of the world. They would lose fortunes in dry wells.

The GCC went too far that day. Although I continued to debate in public with BP's lead person at the negotiations on opposing sides of the issue, he told me privately that he and his employers were embarrassed in the extreme by the GCC's tactics. From 1995, he and others in BP argued increasingly within the company that it should switch sides on the global-warming issue.

In March 1995, at the Berlin Climate Summit, *Der Spiegel* reporters investigated Don Pearlman and trapped "the high priest of the Carbon Club", as they called him, in a lie. A Dutch climatologist told them about tampering Pearlman had organized, via the Kuwaitis, in the IPCC process. At a critical meeting, the Kuwaitis evidently tried to submit amendments, in Pearlman's own handwriting, of otherwise undisputed statements. And at a vital late-night session of talks in New York in February, the Carbon Club had so blatantly ferried instructions to the OPEC delegations that UN officials had told the lobbyists to quit the negotiating chamber. Pearlman denied to *Der Spiegel* that such a thing had happened. A UN official confirmed it, on the record.

Texaco vied with Exxon for the title of worst oil company in the Carbon Club. Its Head of Communications, Clem Malin, conjured up a superb gambit. He somehow got himself appointed as head of the

International Chamber of Commerce delegation to the Berlin Climate Summit, where he was in a position to speak for *all* industry sectors. This, of course, included many particularly threatened by climate change, such as insurance, water, agriculture, tourism, fisheries and the medical industry. As I sat next to him on a debating platform trying not to choke on bile, Malin boasted to a huge audience of media and delegates about the scale of his company's capital expenditure on oil, saying "We are not part of the problem, we are part of the solution."

In November 1995, at the final plenary of the IPCC scientific working group in Madrid, as the crucial policymakers' summary of the Second Scientific Assessment Report was being drafted, that same Don Pearlman – a non-scientist – overtly issued instructions to oil-ministry officials from Saudi Arabia and Kuwait (also non-scientists). So blatant was the manipulation that one senior US government climatologist asked if he could have his name removed from the final report.

In June 1996, the GCC orchestrated a campaign to discredit Ben Santer, a lead author of the Second Scientific Assessment Report. Santer had altered the text of the draft to reflect changes agreed in Madrid, as requested by the meeting. The GCC sought to cast this as scientific fraud, saying "… the changes quite clearly have the obvious political purpose of cleansing the underlying scientific report". The IPCC's leadership pointed out that Santer was merely following agreed procedures.

In October 1997, growing desperate as the Kyoto Climate Summit approached, Exxon was prominent in a US Chamber of Commerce campaign to derail the Kyoto Protocol by casting it as a document that let the developing world off any commitments on greenhouse-gas emissions. Meanwhile, Exxon's boss Lee Raymond told the Chinese at the World Petroleum Congress in Beijing that attempts to curtail fossil-fuel use were "neither prudent nor practical".

Despite this sustained effort to torpedo the negotiations, the Kyoto Summit of December 1997 added a little bite to the enfeebled Climate Convention as negotiated in Rio in 1992. Along the way, the Carbon Club had begun to splinter. BP quit in 1997, finally too embarrassed at the GCC's crude tactics. In *The Carbon War*, I argue that this was one of the main developments that allowed the final negotiation of the Kyoto Protocol. The progressive voices within BP deserve great credit for this. It is a big mistake to think these companies are monoliths of intractable defence of the ruinous status quo. Shell and, to widespread surprise, Texaco also later quit the Coalition.[193]

THE FINANCIAL SERVICES INDUSTRY IN THE RUN-UP TO KYOTO (1991–1997)

The insurance industry, taking over $2 trillion in annual premiums, is bigger than the oil business. It is second only to tourism in income terms. As we saw in Chapter 5, it is gravely threatened by climate change, as is the wider financial services industry. To recap, much of the insurance industry's income is invested, but several hundred billion dollars are retained for property catastrophe losses, which mainly involve earthquakes and climatic disasters. Though losses in recent years have been unprecedented, they have not exceeded a quarter of the reserve pot in any one year. But, in a warming world, disasters are likely to be more numerous, and more intense. Even on current trends, assuming the industry continues successfully to dodge the bullets that are major catastrophes hitting cities, climate expert Andrew Dlugolecki has warned that, in a world doing nothing about greenhouse-gas emissions, net wealth destruction would exceed net wealth creation by 2065 or thereabouts. The danger can be "hedged" to a degree by clever financiers, but there can be no escape from a worst-case assault by nature gone awry.

What has the insurance industry done about this threat to its profitability and indeed its very survival? Virtually nothing. A few representatives have dropped into the international climate negotiations since 1995 for a day at a time. Not a single full-time lobbyist has been deployed. The fossil-fuel industries deploy hundreds. Some insurance companies joined an initiative set up by the United Nations Environment Programme but, despite UNEP's best efforts, it has become a talking shop where meaningful action is far from the agenda. A very few companies have instigated unilateral initiatives – Swiss Re has built a small portfolio of clean energy investments, for instance – but with these and a few other exceptions, the industry is asleep at the wheel of a juggernaut accelerating towards a cliff edge.

The insurance industry is at its most passive when it comes to investment. Most of the climate-risk whistleblowers come from underwriting departments. They are people who understand risk. The investment departments, meanwhile, behave as though global warming has no price implications whatsoever. Much of their vast income they invest in energy, and almost all the energy investments they make are in fossil fuels – the main source of greenhouse-gas emissions. In other words, the insurance industry is channelling a vast river of capital at the very industries, companies and technologies that fuel the global-warming threat to its continued existence. The polite word for this behaviour is dysfunctional.

But then there are proponents, at the very top of the financial services industry, who point to an even a wider dysfunctionality in financial markets.

OIL AND THE ASIAN FINANCIAL CRISIS (1997)

The oil price fell throughout the early and mid-1990s, and the stock markets in the US and Europe rose to record levels. Corporate profits were at their highest-ever level in 1997 but, beyond the borders of the US and Europe, the economic news was not so good. In 1997 banks began failing across all the so-called Asian Tiger economies. By the end of the year, the currencies in Thailand, the Philippines, Malaysia, Indonesia and South Korea had fallen 80 percent against the dollar. A profound credit crunch swept across Asia. Even in once-mighty Japan, weeping bank chiefs announced to the world's TV cameras that they were bust. More than a trillion dollars was wiped out in the storm. Not surprisingly, there was more than a whiff of concern in the air even on Wall Street, where a long-running bull market was in full swing. Little surprise, then, that a small peak appeared in the price of oil (peak 4 in Figure 1, p. 26).

How did such an economic disaster come to pass? The financial pages were filled with complex analyses involving interlocking reasons. George Soros, though, believed that on one level the explanation was simple. He wrote a book about it, and called it *The Crisis of Global Capitalism*. The global capital markets, this living success story of global capitalism argued, are intrinsically dysfunctional. There is no such thing, he says, as the perfect self-correcting market that so many market fundamentalists seem to believe in. Soros warned that the Asian crisis of 1997 might lead to a global depression. It nearly happened, he argued, in the global debt crises of 1981 and 1994. This time, the US and Europe could go under. Where banks fail, credit evaporates, stock markets crash, and economies collapse. This was the reality across Asia in 1997. Collectively the now inappropriately named tigers owed the world's largest banks $400 billion. Suppose they were to default? Under Bank for International Settlement rules, all banks are supposed to keep $8 in capital for every $100 loaned. On that basis, $400 billion of capital could prop up $5 trillion in loans. If the banks had to try and call that in, the world economy would implode. Sensible financial commentators were saying in newspapers that it was a possibility in 1997.[194]

This was all the more likely because international finance had been revolutionized in the digital age. Formerly closed economies had cast off controls and embraced foreign funds at the same time that better technology and financial innovation made it easy to move money instantaneously. But vast inflows can quickly become huge outflows. And financial crises can spread overnight between apparently unconnected markets. The five worst-affected Asian economies (South Korea, Indonesia, Thailand, Malaysia and the Philippines) received $93 billion of private capital flows in 1996. In 1997 they saw an outflow of $12 billion. This shift of $105 billion in one year was the equivalent of 11 percent of their combined GDP.

As Soros said, if not this time, then some time soon. If his analysis is correct, then the financial markets are dysfunctionally unstable even without the prospect of a third oil shock.

WHAT A WAY TO START A NEW CENTURY: OIL AND DOT.COM (APRIL 2000)

And so the first century of oil and gas addiction came to an end. It did so with a clear casebook of history, for those with eyes to see it, of how unstable that addiction had rendered the affairs of nations and economies. It did so with stark evidence of how fragile the world economy had become, of how close nations were sailing to ruin, even without the potentially seismic shocks that could be dealt if the addicts' oil supply dropped by just a few percent for just a short time. And yet it ended with the oil industry and its support structure more powerful than it had ever been: full of people in companies, governments and institutions who imagined that a second hydrocarbon century was not just feasible but probable.

I saw the new century in on a bridge over the River Thames in London, a distressingly empty bottle of champagne clutched in my hand. A thousand years of history was coming to an end. Midnight approached, and shuffling platoons of strangers edged down the streets towards the river, endeavouring to be festive. A few miles along the Thames, the nation's leaders and glitterati were celebrating ten centuries of supposed British greatness in a colossally expensive spiny tent, choosing to forget that we Brits spent much of the first half of the millennium rearranging the marbled palaces of Roman civilization into crude huts. A wall of fire was supposed to race down the river to signify the century's end. How apposite, I thought. A gratuitous and unintentionally symbolistic hydrocarbon burnfest. I looked around me, wondering how many of the throng might be thinking the same kind of thoughts as me.

By amazing coincidence, the Dow Jones reached its highest value ever on the last day of the twentieth century. The first few days of the twenty-first century saw a wobble, but the bull market roared on into March. Business magazines exhorted people to get wise to the wonderful opportunities of the dot.com era. The "new economy", they called it. "The bottom line", according to *Business Week* in the second week of March, was this: "... if you're not heavily investing in tech these days, you're out of luck."

Venture capitalists figured that, of $10 billion going to Europe as a whole in 2000, $3.5 billion would go into British internet companies. Fifty of the latter had launched initial public offerings, with many more expected. Big companies making hundreds of millions were driven from the FTSE 100 index by tiny newcomers who had yet to make a penny. Those ousted included major energy companies like PowerGen, with millions of real customers and guaranteed billing measured in the billions. Such companies were being dismissed as "old economy" stocks by the business magazines at the time.

Others were beginning to think very differently. One eminent economics editor viewed the whole thing as potentially analogous to an investment craze in the 1630s, when Dutch investors went mad for tulips, chasing the price of a single bulb up to $150,000 in today's prices. That bubble burst with a vengeance, ruining many who still held stock. The chain of greed could be extended no further, there being no more "greater fools" left for speculators to sell stock on to at a profit.

Entering March, the price of oil reached $30 a barrel, the highest it had been since the Gulf War ten years before. The *Business Week* headline read: "It's still too early to panic at the pump". Others disagreed. The three post-war recessions had all happened after the price of oil exceeded $22. The year 2000 price rise – a trebling in twelve months – was as rapid as the 1979 one. Yet even at $30 for a 27-gallon barrel – including the costs of transporting the stuff round the world – oil worked out at a mere 13 cents a pint.

A high oil price and clearly puffed-up valuations for companies with no proven profit-making capacity was too much finally for investors. Once again we got a smell of what an intrinsically fragile economy and a sky-high oil price is capable of. On 14 April 2000 Wall Street took its biggest one-day fall in history. The $2 trillion nosedive, equivalent in magnitude to losing the entire current German economy, affected everything: technology-based Nasdaq, blue-chip Dow Jones and broad-based S&P500 all had their biggest-ever falls. Could this be it? I thought. Would it be the beginning of a global recession, or worse, triggered by the irrational exuberance of herd-mentality investors besotted by a ridiculous dot.com model, coinciding with a high oil price?

The figures were mind boggling. Although it was perhaps most worrying that the blue-chip "old economy" stocks fell so far, technology took the biggest hit. Nasdaq lost a quarter of its value – almost all the value accreted since October 1999, with a 25 percent fall in one week, 10 percent in just one day. Microsoft lost $239 billion, the equivalent of the Belgian GDP; Cisco $167 billion, the equivalent of Poland's GDP; Intel $100 billion, the equivalent of Ireland's GDP. An analyst quoted on the front page of the *Financial Times* said: "We had greed, now we have fear."

That's what the first phase of a great global energy crisis would feel like.

And consider this by way of comparison, as you digest the figures. As we saw in Chapter 4, Goldman Sachs – whose oil analysts in 2004 do not believe there is potential for an early oil topping point – think that oil demand cannot be met by supply unless the oil industry invests something approaching $250 billion per year in new infrastructure and exploration every year for a decade.

We can get a further feeling for the potential impact of the early topping point, and how we have allowed society to become so vulnerable to it, in the strange tale of the UK "fuel crisis" of 2000.

THE UK "FUEL CRISIS" (OCTOBER–NOVEMBER 2000)

After the dot.com crash, the global economy moved into a downturn that lasted several years. The price of oil stayed high. Eighteen months before, the real price of crude oil had been its lowest in post-war history. For a time in 1999 it was actually one half of the real price of oil in the 1950s, and one fifth of the real oil price at the beginning of the 1980s. Entering September 2000, it reached a ten-year high of $34 a barrel. It had trebled in just over a year. $40 a barrel could not be ruled out, analysts said.

In Washington, Bill Clinton raised the prospect of recession as a result of the high oil price with a Saudi Crown Prince. The Saudi government promised to boost production to drive the price down. Such promises were soon to become a regular feature of the new century.

Sheik Yamani spoke out in his retirement. He told Reuters that OPEC was behaving in a way guaranteed to accelerate the end of the oil era. "OPEC has a very short memory," he counselled. "It will pay a heavy price for not acting in 1999 to control oil prices. Now it is too late. The Stone Age came to an end not for a lack of stones and the oil age will end, but not for the lack of oil." He feared the West would go all out for fuel-cell and other non-gasoline engines. "Technology is the real enemy for OPEC," he said. "The real victims will be Saudi Arabia with huge reserves they can do nothing with – the oil will stay in the ground for ever."

All this news of rising prices did not sound good to, among others, the truckers of Europe. Fuel-price protests began almost at once, initially in France. Hauliers seeking fuel-tax cuts blockaded fuel depots, and 80 percent of the country's petrol stations soon ran dry. Europe, and the UK in particular, was about to be dealt a savage reminder of what a high oil price can do to economies as fragile as we have allowed them to become as a result of our oil addiction, in concert with some of the other intrinsic weaknesses George Soros speaks about.

The protests soon spread to the UK and Germany. Farmers and truckers staged a blockade of a Shell refinery at Stanlow in Cheshire, and within days – despite tiny numbers of blockaders – more than a hundred British filling stations ran out of fuel. Fifteen main refineries in the UK supplied 80 percent of the fuel used at the time.

OPEC ministers met in Vienna and agreed to increase their production by 3 percent in an effort to bring prices down. But the

European truckers seemed to be more annoyed with their own governments' high-tax policies than the producing countries or the oil companies.

By 13 September 2000, 90 percent of UK filling stations had run out of petrol. The country was two days away from food shortages. Panic buying had started at supermarkets. Buses and trains were cancelled nationwide. Schools shut down. Hospitals had to cancel operations. Even for political junkies like me, long used to the speed with which political events can leap from action to reaction, it was stunning to see the stark evidence of how dependent we had let ourselves become on oil and, more than that, just-in-time delivery of oil.

Prime Minister Blair spent a whole afternoon on the phone to the bosses of the big oil companies, asking them to tell their tanker drivers to roll, guaranteeing that the police would protect them from the few angry truckers. Press reports revealed that government ministers were privately fuming, saying that the oil companies were deliberately sitting on their hands. A Home Office adviser, Lord Mackenzie – a former senior police officer – said, "Roads are often not blocked. If they were, the police would not hesitate in arresting people. It seems to be the companies who are causing the difficulty, and there does seem to be some collusion with the protesters. Perhaps it's because if the fuel duty was reduced they would sell more petrol."[195]

On top of police action, the companies could use the law to remove protesters from their depots and refineries. They hadn't done so. Instead of serving court injunctions on truckers, Shell was handing them hot coffee and refreshments. "They are right behind us," one haulier told the *Financial Times*.

The Prime Minister went beyond phone calls. He asked nine oil company bosses in to Downing Street to exhort them to get their tankers on the road. To this meeting with a head of state leading one of the world's largest economies, Shell sent – as you would expect – their UK Chief Executive, Malcolm Brinded. BP, which had recently rebranded itself as Beyond Petroleum, sent a mere vice president. After the meeting, the oilmen read out a statement on the steps of Number Ten. It fell well short of an assurance that their tankers would start moving again. They said they still had fears for the safety of their drivers. Yet the worst recorded incident of violence had been someone throwing a plastic traffic cone against the side of a tanker.

Stock markets, meanwhile, were slipping predictably. The long petrol queues were unsettling dealers. One told the *FT*: "These are things that everyone thought were consigned to the past." The OPEC decision to hike production was having no immediate effect.

Entering week three of the crisis, the health service was on red alert. Supermarkets had begun rationing bread. Companies, unable to operate production lines, were sending workers home. The Prime Minister, thinking he had talked the oil companies into action, and having told the nation then that the problem was twenty-four hours from resolution, now admitted that we faced a national crisis without an end in sight. He filled eighty military tankers with Ministry of Defence petrol in readiness to cover essential services.

Tanker drivers for both BP and Shell contacted the newspapers to say they had been explicitly ordered not to go through demonstration lines, even when police had cleared the streets. Beyond Petroleum had had an opportunity to spin the crisis into an argument for going "beyond", but it had chosen to join Shell and the others in spinning an opportunistic defence of the status quo.

Government spokespeople began hinting at reprisals. Great, I thought. Go for it, please. Slap windfall taxes on them and put the revenue into commercializing the technologies that Sheik Yamani worries about. The ones that would truly take us Beyond Petroleum.

After a disappointingly long silence, the environment groups at last joined the debate. A letter in the *Guardian* from the Director of Friends of the Earth extolled the need for high fuel taxes, noted that the government was sitting on £4 billion of unexpected oil-tax revenues because of the high oil price, and urged them to spend it on public transport. Nowhere else, in the entire crisis so far, had I seen or heard any other reference to the environmental imperative for petrol to be taxed.

On 15 September, the Chancellor announced that he wouldn't give in to the fuel-tax protesters, even though polls were showing that the protests commanded huge public support. But the protesters were facing a barrage of calls to back down, even from their supporters. Nobody wanted to see the country actually brought to economic ruin. The populist *Daily Mail* exhorted the protesters to call off their flimsy blockades, making it clear that it supported their campaign and would continue to support it. The truckers' leaders said they would call off the dogs for a while, but would restart the blockades in a few weeks if the government had not cut fuel tax by then. The tankers began to move again.

Analysis of the crisis started, and finally the environment appeared in the media coverage. "Do puny governments hold sway any longer over these global energy corporations?" an editorial asked. "The environmental project that flowered in Rio and Kyoto is parched. Getting ordinary people to change their daily behaviour for the sake of the planet now looks more difficult than ever." The cartoon of the day showed a fat trucker driving a tanker with all the oil company logos splashed on it, including BP's sunflower. A cat, even fatter and dressed in a suit, sat with its arm round the driver, its paw stuck out of the window, and a single claw stuck in the air. A miniature prime minister rode on the bumper of the tanker as a mascot.

By 20 September I could bear it no longer. My company took out a full-page advertisement in a national newspaper. "Call that a fuel crisis?", shouted our message in letters two inches tall. "You thought last week was bad", the text read. "If we keep relying on fossil fuels the resulting global warming will kill more people than all the twentieth century's wars put together. Britain's only hope is to invest in sustainable forms of energy. Something the governments of Germany and Japan had already recognized. To date our environmental investment had been almost non-existent. Unless we did something soon it was clear we would have more than the price of petrol to worry about." It went on to invite the reader to find out more about one alternative, solar power.

The government, temporarily reprieved, sat down to work out how to deal with the angry truckers and their powerful supporters. They had only a few weeks to work it out. The weather provided the solution for them. On the night of 30 October, it was very difficult to sleep in the UK. Wind and rain lashed windows with a rare ferocity. I arose the next morning to find a country once again on its knees. Dozens of rivers had broken their banks, thousands of trees were down and almost no trains were running. Hundreds of homes were flooded. People were wading through their lounges amid human sewage that had spilt from drains. Lifeboats were rescuing stranded people thirty miles from the sea.

The front-page headline in the *Guardian* the next day read "Global warming: it's with us now". The news was replete with references to climate change, and government ministers were among those running the line. Insurers calculated that the storm was the third-worst insurance loss ever in the UK, with early estimates at £500 million, and they too blamed global warming.

In the first week of November, the connection between floods and the greenhouse effect was being made everywhere. The floods were the worst for fifty years, and it seemed there was worse to come, with large rivers like the Severn taking a few days to reach peak levels, and more bad weather forecast. Estimates of the final insurance bill passed a billion pounds. The Prime Minister visited flooded Worcestershire, looking appropriately worried. "We have to work to deal with the issue of climate change," he said. Even the *Sun* newspaper, a tabloid with a long history of belittling matters environmental, mentioned global warming. The worst-case analysis was out in the open too. Previewing the annual climate summit, this time to be held in The Hague, a long article in the *Financial Times* mentioned the "remote but real" possibility that warming could trigger a runaway greenhouse effect, in which the warming stimulated by fossil-fuel burning would trigger such vast releases of carbon from drying soils, dying forests, stratifying oceans and melting permafrost that they could not be stopped even if all fossil-fuel burning ceased.

Another storm was expected. Satellite photos of it appeared on the front pages of newspapers. The British were no longer just talking about the weather, they were living in fear of it. After the storm, with eighty-one flood warnings in place across the country, the cartoonists had a field day. One showed a group of placard-carrying truckers up to their knees in water demonstrating in favour of cheaper global warming. God's giant arm reaches down to them from a dark rain cloud, a piece of paper in his hand. Written on it: "The Bill".

Encouraged by the potential turning of the tide, so to speak, the government shaped up to tough it out with the truckers on their ultimatum. They had no intention of dropping lucrative fuel taxes, and mobilized the army to drive oil tankers if the truckers disrupted the refineries. The truckers, for their part, threatened that 25,000 crawling lorries would converge on London if the government didn't cut fuel tax by 26 pence a litre.

By now, the 10 percent of the UK population that lived on flood plains did so in misery. Most of the UK's major rivers, notably the Severn, the Ouse and the Derwent, had made extensive use of these flood plains. Thousands of homes were flooded. The sandbags were out even along Millionaires' Bank on the Thames. A toxic waste dump was under threat. Major railway lines were still closed.

An opinion poll showed that 78 percent of British people now thought that their families would be affected by climate change over the next few decades. But it also showed widespread ignorance about what was causing it. Only two-thirds thought cars contributed. Less than one in five associated greenhouse gases with domestic appliances.

The Chancellor lost his nerve at the eleventh hour and knocked 8 pence off a litre of petrol. It was not enough for the protesters, and the second round of the fuel protests went ahead on 10 November. They were a complete flop. Instead of 25,000 lorries, a maximum of forty headed for London, massively outnumbered by police. Environmental protesters in favour of higher fuel taxes were there this time.

"People who want lower fuel taxes are like turkeys campaigning for Christmas," a campaigner told the press.

16 May 2001, London
Three hundred of BP's young high-flyers, flown in from all corners of the world, sit in a City of London hotel, all dressed in blue polo shirts, at round tables for ten. Forty vice presidents sit scattered among them. The company bills this event as their global forum. At it, "the leaders of today meet the leaders of tomorrow". I am invited to join them, the only speaker in the forum from outside the company.

I stand drinking coffee before my session. One of the VPs walks over to introduce himself, a big smile on his face. In the late 1970s, Tony Hayward – now Group Vice President for finance, one of only four reporting to the Chief Finance Officer and viewed as a potential successor to CEO Lord Browne – was a long-haired graduate student in geology at Edinburgh at the time I was an even longer-haired youthful lecturer at Imperial College. I knew him well. We both worked on the geology of ancient oceans. But I haven't seen him for more than a quarter of a century.

"Tony," I say, "you've had a haircut."

Tony, jocular now as I remember him to be then, wants to know how I think I can run a business. Good question. I want to know how he has learned about money and oil, there being not a drop of it in ancient oceans. We have an entertaining half hour before I have to perform.

I call my presentation "BP and the world by the time you are on the board". I have designed it, so I hope, to hit hard but not to alienate.

At the end, the organizers conduct some on-the-spot opinion polling. Everyone in the audience has been issued with a hand-held device that can register anonymous votes on questions answerable with a choice of six categories: strongly agree, agree, neither agree nor disagree, disagree, disagree strongly, or undecided. The results feed into a computer that calculates percentages and instantly relays them to a giant screen.

The organizers test the system by asking: "Are you male or female?" Fully 3 percent of BP's finest profess to be undecided.

I have been requested, ahead of the event, to supply five questions. I now run through them.

First I ask: "Do you consider you are sufficiently well briefed to have a considered opinion about global warming?" This is a necessary precursor to the four following questions, and – in part being feedback on the presentation itself – a truth game with humiliation potential. A sweaty palms moment then.

66 percent agree, 18 percent are undecided, and only 16 percent disagree.

Not bad, I think.

Next I ask: "Does global warming pose a serious threat?"

85 percent agree, 10 percent are undecided, and only 5 percent disagree.

I thought I might encounter majority agreement, but nothing like this.

Then I ask: "Does global warming pose a serious threat to BP within the next ten years?"

40 percent agree, 28 percent are undecided, 32 percent disagree.

A majority, albeit not a big one, on a timescale well within their vocational lifetimes. Fascinating.

I then pose two questions that I hope will prove subversive. First I ask: "Can BP lead the micropower revolution?" I make clear what I mean here. The question is one of in-principle technical capability, not real-world practicality. "Micropower revolution" I define in the sense that the Economist *and the Worldwatch Institute, for example, use the term: it refers to the full family of renewable energy technologies minus large-scale hydropower, plus fuel cells and microturbines, including natural-gas-fuelled combined heat and power.*

60 percent agree, 19 percent are undecided, and only 12 percent disagree. [I must have copied this down incorrectly, since it only adds up to 91 percent. The most likely mistake is 69 percent not 60 percent.]

Then comes the final question, the one I am most interested in: "Should BP kick-start the micropower revolution?" I make it clear that this is the ethical question. If they accept the premise that global warming is a major problem, and if they think that BP has the tools to lead the fight against it, do they – the BP leaders of tomorrow – think that the company has a moral imperative to do the leading?

68 percent agree, 14 percent are undecided, and only 18 percent disagree.

23 percent – almost a quarter – agree strongly.

I return to my office flabbergasted.

John Browne was not among the leaders of today dispersed in the audience as I spoke. He had addressed the forum in the morning and departed. Two days later I wrote to him to share my flabbergastedness, and what I saw as its implications for his company and his role as current leader of it. I knew it was an impolitic thing to do. I knew it would go down like a lead balloon with the organizers. But George W. Bush had just announced his travesty of an energy plan. He planned to build a new power plant every few days for the next twenty years. A quarter of a million miles of new gas pipelines would be built. The Arctic National Wildlife Refuge would be opened up for oil drilling. The Clean Air Act would be revised to allow more pollution from power plants. And then there were the things missing, like any commitment to improve vehicle fuel efficiency. Sixteen of the world's national academies of science had just issued a statement on global warming, directing themselves at America with perfect timing. The IPCC is "the world's most reliable source of information on climate change", they said. They endorse its conclusions. "Despite increasing consensus on the science underpinning predictions of global climate change, doubts have been expressed recently about the need to mitigate the risks posed by global climate change. We do not consider such doubts justified." So I just couldn't stop myself writing to Browne.

"That nearly seven in ten of your leaders of tomorrow would seem to be looking for BP to lead a process to decarbonize global energy supply today stunned me," I offered. "There can be no comfort in the view that they think the company is already doing enough, as things stand. A glance at my presentation will show that the analysis I offered on the current position on global warming and energy policy in the group was very downbeat. As you can imagine, there followed a fascinating Q&A session. In this nobody took issue with the arguments presented, or tried to argue that BP is doing anything more than the 'useful tokenism' I described in the presentation.

"It seems to me that the corollary of my session at the Global Forum could not be more vital for your strategic thinking about sustainability. The conclusion has to be that BP's immediate leadership will probably not be able to hold the company's internal constituency firm behind a future where you increase oil exploration without aggressively leading the world in commercializing micropower.

"One young man asked me, in front of his peer group, what I would advise a young petroleum engineer to do, having heard such a lecture. I answered in essence: stay on and help BP change from within. But the poll suggests they won't stay, and they certainly won't stay happy, loyal and as brilliant as you need them to be unless the company does much more than it is at present to lead the way to global greenhouse-gas emissions reductions."

I probably should have left it at that, but I just wasn't able to.

"One questioner asked me how I thought the current board and investors would have voted on the questions I posed," I continued. "I answered that I had no idea, but that I would love to know. I wonder, therefore, if there is any possibility you might be able to offer me the opportunity to present the ideas in the lecture to board members and key investors?

"It is becoming increasingly probable that governments cannot solve the dire problem of global warming now, so huge has it become after a decade of delay in responding. It seems clear to me that industry must lead the way by change from within. If BP cannot lead that process, I asked the forum, then who can? I do not know the answer to that. If the answer is nobody, then whither hope for humankind? And whither hope for the young people at your forum? Nobody can do business in a world collapsing under environmental and economic pressures one large group of intergovernmental experts described as 'second only to nuclear war', should we be collectively blind enough to maintain course with oil, gas and coal dependence."

As I deliberated the wisdom of committing this rant to the postbox, an e-mail came in from the organizer of the forum, Aleem Sheik. "There was a lot of positive reference to you throughout the event," he wrote, "and you captured many minds, including Group Vice Presidents."

I posted the letter.

A week later a reply from Browne arrived on my desk. He did not seem to agree with Mr Sheik. After a few pleasantries, he came straight to his point.

"I am sorry that you were not present to hear the Group Vice President of Technology, Andrew McKenzie, who spoke after you, about all that BP is doing in response to global warming. I believe that might have changed your views considerably, not only in enhancing your understanding of the leading role that BP is playing in the area of global warming but also in the assessment you made of the opinions of our young graduates. I am convinced that these leaders of tomorrow will play a key role in ensuring that BP remains at the forefront in all its dimensions."

THE BUSH YEARS AND THE SECOND GULF WAR (2000–?)

George W. Bush entered office in January 2001. The same month, the Global Climate Coalition dissolved itself. As a Greenpeace spokesman saw it, "the last stronghold to prevent progress on global warming" had gone. "Now President Bush and Exxon stand alone." I didn't agree. "We have achieved what we wanted to accomplish with the Kyoto Protocol," the Coalition's spokesman said.[196] That sounded more like it to me. They simply weren't needed any more. He might have added that they had champions in the White House from the top down, too, not to mention many in Congress. President Bush and Exxon hardly stood alone.

Vice President Cheney, the former CEO of giant oil-services firm Halliburton, took control of the Bush Administration's energy plan. The plan de-emphasized efficiency and alternatives, and pulled out every stop to scrabble for the remaining oil, at home and overseas. It actively chose oil dependency. A key talisman was and is the Arctic National Wildlife Refuge, where relatively big oilfields lie unexploited. The Bush Administration is set on seeing the 10 billion barrels or so that supposedly lie there produced. Given the time it would take to extract and transport the oil from such frozen, inhospitable and far-off terrain, the best they can hope for is six months of American supply in ten years' time.[197]

The lessons of the past oil shocks seem to have sailed past the consciousness of government, the oil industry and the auto industry alike. The US Congress established fuel-efficiency standards in 1975 in response to the oil embargo of 1973. At the time of passing the Corporate Average Fuel Economy (CAFE) laws, light trucks were allowed to meet a lower fuel economy standard because they constituted only 20 percent of the vehicle market and were used primarily as work vehicles. Today, light trucks comprise nearly 50 percent of the new-vehicle market, and are used primarily as passenger cars. Fuel-economy standards for cars are now 27.5 miles per gallon (mpg), and for light trucks 20.7 mpg. In March 2002, senators John Kerry and John McCain introduced an amendment to the Energy Bill that would have increased fuel-economy standards for cars and light trucks to 36 mpg by 2015. This would have saved 2 million barrels of oil a day by 2020. After strong lobbying from General Motors and Ford, the Senate rejected it.[198]

Perhaps even more worrying than the willingness of Congress to bow to the wishes of Houston and Detroit is the evidence of how oil money returning from Saudi Arabia has inveigled the fabric of modern America. Over the last three decades, American consumers have transferred a staggering $7 trillion to OPEC producers.[199] According to Craig Unger in his book *House of Bush, House of Saud*, since the mid-1970s 85,000 super-rich Saudis have invested fully $860 billion in American companies. In among this, he estimates that $1,476 million has made its way to companies and institutions allied to the Bush family.[200] At best, this has to deepen the difficulty of breaking the cycle of oil dependency. At worst … well, I'll leave all that to Michael Moore. The point is this. The deeper the dependency on oil and oil money becomes, the worse the effects of the unforeseen energy crisis will be when it hits, so the more America's security is undermined, even as its government advances enhanced security as the rationale for the latest actions of the Pentagon's global oil protection service.

Then there is the terrorist, as opposed to economic, threat to America's security. The al-Qaida network was forged in the Afghan War, where Saudi volunteers including Osama bin-Ladin went to fight

Russian invaders armed with weapons financed by both their own government and by America. A common liberal view of twentieth-century American history is that the US has tended to create its own enemies. It is difficult to think of a better example. It was American money that helped arm and train bin-Ladin's cadres, at least originally. It was American presence on Saudi soil after the first Gulf War that tipped his hatreds into terrorism. Having defeated one infidel in a neighbouring country, he certainly didn't want to see another quietly invading – as he saw the American military presence after the defeat of Saddam Hussein – his own back yard. So he began the plotting for the event that would change the world, 9/11. Following al-Qaida's repulsive act that day, America has long since lost the well of international goodwill and sympathy it triggered. At the time of writing, as Muslims fall to American bombs by the day in the Middle East, bin-Ladin couldn't have had a better global recruitment plan if he'd designed it himself.

If America hadn't become so dependent on Middle Eastern oil, there is every chance that none of it would have happened.

2004: $50-A-BARREL OIL AND RISING

The oil price began to climb with a vengeance in 2004. The sequence of major events, and how people and institutions reacted to them, suggest much to us about the not-too-distant future. Mixing major developments on global warming into this account of very recent history is also instructive, in that viewing the two together gives us a feel for how they might conflate in the coming energy crisis.

Winter 2004: discontent and lies

The year 2004 opened with a blasé announcement by Shell in January that it might have overstated its reserves by 20 percent, or some 3.9

billion barrels. The Chairman of the Committee of Managing Directors, Sir Philip Watts, did not even make the announcement himself. Major shareholders immediately called for his resignation.[201]

Meanwhile, the UK government's Chief Scientific Adviser, Sir David King, was also making headlines by attacking the US over global warming. "In my view, climate change is the most serious problem we are facing today," he reasoned, "more serious even than the threat of terrorism."[202] Later that month, it emerged that the Pentagon had been busy researching the same line of thought. An internal study found its way to *Fortune* magazine. It spoke of a plausible scenario that involved a total or partial shutdown of the ocean conveyor leading to horrifically hard winters, violent storms and droughts. As for the outcomes of these and other coalescing impacts, "as the planet's carrying capacity shrinks, an ancient pattern re-emerges: the eruption of desperate, all-out wars over food, water, and energy supplies."[203]

With this kind of attention to the greenhouse issue building up, the oil companies went into their 2004 AGMs in March facing a record number of shareholder resolutions on global warming. Most of these questioned how the companies would respond to emerging regulatory pressures on emissions, or sought disclosure of emissions data. ExxonMobil and ChevronTexaco were asked for a report on their efforts to invest in renewable energy. "The disparity of preparedness among the companies is disturbing," said Ceres analyst Andrew Logan. "All oil companies essentially operate in the same global markets and are susceptible to the same emerging regulatory structures around the world – yet many of these companies seem relatively uninformed about the issue and how it could affect prices. It seems that US intransigence on global warming has translated into insularity that puts US companies at serious risk."[204]

Shell, meanwhile, downgraded its reserves again, to further howls of anguish from investors. America's mass-tort lawyers began to take an interest. "Shell holed below the waterline," read a typical headline.

BP inevitably faced questions about its own reserves, but CEO Lord Browne brushed them aside. In March, he went to Washington, where Matthew Simmons had spoken out about his fears concerning Saudi oil supplies in February, to address the US National Press Club. "There is no physical shortage," he said categorically. "The resources are there. The world holds some 1,000 billion barrels of oil which has been found but not yet produced, and some 5,500 trillion cubic feet of natural gas – also found but not yet produced. At current consumption rates that is forty years of oil supply and sixty years of gas. In addition the US Geological Survey estimates that some 800 billion barrels of oil and 4,500 trillion cubic feet of natural gas are yet to be found."[205]

Shell parted company with Sir Philip Watts. On 20 April it became clear why. The company made its third downgrade, this time admitting that it had misled investors. Devastating e-mails showed that Watts and Walter van de Vijver, Head of Exploration, had known about the problem for at least two years and possibly as long as seven. One dated 9 November 2003, from van de Vijver to Watts, read: "I am becoming sick and tired of lying about the extent of our reserves issues and the downward revisions that need to be done because of far too aggressive/ optimistic bookings." The two executives had insisted that the problem only came to light late in 2003. But one of the e-mails showed van de Vijver knew that SEC rules were being broken in February 2002. As recently as May 2002 Watts had told van de Vijver that he should "leave no stone unturned" in making the figures even higher.[206]

Spring 2004: Iraq goes pear-shaped

Meanwhile, the war in Iraq was entering its second year, with proliferating insurgency, increasingly beleaguered US and UK forces, and attacks on oil pipelines so commonplace that daily operation of Iraqi oil infrastructure, much less upgrading and expansion of production, was becoming extremely difficult. The Abu Ghraib torture scandal broke in May, further deepening the mire.[207] BP and Shell reacted to this in differing ways. In the last week of April, Browne announced he would be pulling BP out, citing security and political pressures. This dealt a catastrophic blow to the Bush and Blair plans. Iraq had no chance without the giant companies, because it needed their expertise, technology and cash to get the domestic oil industry running.[208] The following week, Shell announced it intended to establish a "material and enduring presence" in Iraq. It needed fresh reserves, a spokesman told the press.[209]

Unsurprisingly, the oil price started to creep up. In May 2004, it went through the $40-a-barrel mark for the first time. Continuing rises forced US consumers to spend an extra $44 billion at the pump during the first six months of 2004. By June, retailers were feeling the pinch in falling sales as shoppers stayed at home. As one retail executive saw it: "We are hurt by high oil prices because people are giving their extra dollars to Exxon."[210]

In late May, Shell made a fourth reserves' downgrade, just before BP published its Statistical Review, with the conclusion that world reserves had gone up 10 percent during 2003. BP Chief Economist Peter Davies said: "There is no global oil resource or reserve shortage. Oil production continues to be replaced."[211]

Elsewhere, optimism was even more extreme. One Dallas-based analyst told Voice of America that "the estimate is that we have about 14

186

trillion barrels of oil in shale oil and tar sands. Now, that is enough to fuel us for the next 500 years."[212] Another professed that we would never run out, citing Russian theories that oil could be created underground by inorganic means, that is from carbon in no way related to microscopic life in ancient seas, or indeed life of any sort.[213]

Summer 2004: jitters everywhere, $100-a-barrel oil fears

An al-Qaida attack on Saudi oil infrastructure deepened jitters in June, adding a new factor to the list of reasons commonly cited for the persistently high price at the time: high demand, especially in China, shortage of spare capacity, lack of refinery capacity and – increasingly – speculation, with hedge funds betting on a long-term high price, so helping to create a potentially self-fulfilling prophecy.[214] Though it did not cause major disruption, the attack pointed to a bigger possibility: a highly disruptive terrorist act at the main Saudi oil-export terminal at Ras Tanura. As a financial analyst, former Mobil engineer Fadel Gheit, put it: "If you can blow up the Pentagon in broad daylight, then it cannot be impossible to fly a plane into Ras Tanura – and then you are talking $100 [per barrel] oil." Regime change would have the same effect, Gheit suggested. He views replacement of the Saudi royal family by a militant regime, as happened in Iran, as "only a matter of time".[215] The price fell back from $40 in June 2004, but stayed in the high $30s. In Iraq, oil exports were brought to a halt after two strategically located attacks on pipelines. An oil-industry executive was assassinated the same day.[216]

In July 2004 I attended a gathering of oil analysts in the City of London to observe their culture of denial at work. The subject matter was the Shell reserves scandal and the details of how reserves should be reported, but they addressed the bigger question. "We're not running out," scoffed one, and went on to pinch Sheik Yamani's line: "Ultimately, the Petroleum Age will come to an end. The Stone Age came to an end too. But not because we ran out of stones."[217]

A Deutsche Bank analysts' report published a month earlier had offered an insight into the intellectual gymnastics that are required to maintain course with this kind of denial. It opened with a statement of the obvious: "If it could be shown that geological constraints signify an imminent peak in oil and natural-gas production levels, the investment implications, in our view, would be enormous." But it went on to conclude that there was no need for concern, and criticize the use of the Hubbert Curve on the grounds that technology, costs, prices or politics can move the descending part of the curve to the right. "In our view, this completely eliminates its value when trying to make most stock or sector allocation investment decisions," the analysts concluded. Their report spun a spurious attack on the Hubbert Curve for oil depletion, which few early toppers think can be applied rigidly to countries other than the US – as discussed in Chapter 3 – into an assumption that there is no need to look at its general implications for depletion.[218]

Meanwhile, scientists announced that atmospheric carbon-dioxide concentrations had risen by an alarming 3 parts per million – ahead of projections – for the second year running. Though too soon to be sure this was a trend that could indicate amplifying feedbacks running amok, it was enough to push speculation about the dreadful prospect of a runaway greenhouse effect back into the media.

The fate of life as we know it couldn't hope to occupy the same column inches as the Shell scandal, which rolled on like a bizarre soap opera. In late June, investors vented their fury with the "incompetent" board at the company's AGM, having learnt that non-executive directors knew of the downgrade weeks before it was announced.[219] In late July, Shell paid penalties totalling $150 million to the US Securities and

Exchange Commission and the British Financial Services Authority. In the latter's eyes, the world's third-biggest oil company stood accused of "unprecedented misconduct".[220] A number of serving and former directors remained under criminal investigation by the US Justice Department.[221]

The oil companies posted record mid-year profits, but the questions remained about their ability to keep up with demand. At best, in the case of ExxonMobil, production was relatively flat. But many companies produced less than they did in 2003; in Shell's and ConocoPhillips' cases 5 percent less.[222] Meanwhile, BP's Lord Browne travelled to Russia to seek personal assurance from President Putin that his government's legal assault on Russian oil giant Yukos – a thinly disguised renationalization – would not affect BP's growing dependency on Russian oil.

In early August 2004, a Saudi expert told the *Oil & Gas Journal* that high oil prices are inevitable for the rest of the decade.[223] OPEC itself warned about supply problems. Secretary-General Purnomo Yusgiantoro announced that there was no longer any spare capacity: the taps could not be opened further. "The oil price is very high, it's crazy," he said. "There is no additional supply." The oil price moved higher on the back of this indiscretion, nudging the $45 mark.[224] The following day Yusgiantoro tried to calm the market by saying that OPEC could in fact pump another 1 to 1.5 million barrels per day. It was not enough to stop headline writers shouting about a threat to the world economy.[225] The very next day the record was broken again, this time on news of another blow dealt by the Kremlin to beleaguered Yukos.[226] "The risks of a crash have increased, the oil price could be the tipping point," observed a London-based analyst.[227] "It isn't called 'the devil's excrement' for nothing," quipped a newspaper leader column.[228]

The hits rolled on. Iraq was forced to shut production in its southern fields after the threat of sabotage.[229] Deutsche Bank warned for the first time that the oil price could reach $100. "It's worth asking ourselves", said their Global Energy Strategist Adam Sieminski, "what would happen if we lost four million barrels a day, due to some accident? Or let's say Iraq's two million barrels a day became unavailable. OPEC's got no spare capacity. And that could be it: $100 per barrel."[230]

AUTUMN 2004: ALL EYES ON OPEC (AGAIN)

OPEC ministers gathered in Vienna in September to confer about the growing crisis, amid much confusing headline-making by oil-industry chiefs. Total CEO Thierry Desmarest took a break from his scoping of a Shell takeover to announce that OPEC should open their doors – closed in most cases since the 1970s – to the big companies to ensure that supply meets demand. "If you want to increase production capacity, it is key," he said. "We create opportunities through exploration, but exploration takes a lot of time."[231]

What message does this give about expectations for new exploration finds in the non-OPEC world?

Browne responded the next day with a bullish alternative view. "There isn't really a supply crunch at the moment. We have the perception of the risk of a supply interruption, but that's all we've got."[232] He told the *Financial Times* that people were pessimistic about supply meeting demand because of their reluctance to consider politically fraught regions like Russia, west Africa and the Middle East. "When you look at it from the West, the domestic sources of production are definitely in decline. So close in, that could look universally applicable. But of course, it's not. I think oil is coming from places [people] don't understand."[233]

ExxonMobil's Lee Raymond sided with Total. He told the OPEC

ministers that growing international demand would be met only through access to oilfields currently off-limits to the international oil companies. "The future need for petroleum energy will be such that restrictions, in whatever form and wherever imposed, will jeopardize access to adequate energy supplies for world consumers." What did this mean for the US and its oil policy? "I think that the notion in the United States of energy independence, which was first proposed in the Nixon administration, was a poor concept thirty years ago and it is a poor concept today."[234] There it is, just as he told OPEC. It couldn't be clearer. We choose dependency, and therefore overseas adventures by our military in support of our dependency.

OPEC announced its response to building events: a 1-million-barrel hike in production.[235]

Surely, I thought in the summer of 2004, by now governments must be noticing the early depletion problem. I met many a senior official in the course of my work on the UK government's Renewables Advisory Board, and rarely missed an opportunity to ask where they thought the oil peak was. Not a problem, I was told by someone near the top of the Department of Trade and Industry. Don't know anything about it, a Treasury official told me. Nothing in it for politicians, said someone in a high place in the Prime Minister's office. Then, in September, an important letter began circulating in Whitehall. It was from the First Secretary for Energy and Environment in the Washington embassy. He had attended a seminar on oil supply given by the respected consultancy PFC. "The presentation drew some gasps from the assembled energy cogniscenti," the diplomat reported back to London. "They predict a peaking of global supply in the face of high demand by as early as 2015. This will lead to a more regionalized oil market, a key role for West African producers, and continued high and volatile prices."

In October, G7 finance ministers met in Washington for the annual meetings of the World Bank and the International Monetary Fund. A key part of the statement by the ministers and central bank governors read: "Oil prices remain high and are a risk. So first, we call on oil producers to provide adequate supplies to ensure that prices remain moderate. Second, it is important consumer nations increase energy efficiency. Third, it is important for consumers and producers that oil markets function efficiently and we encourage the International Energy Agency to enhance its work on oil data transparency." As a veteran financial correspondent described it, "… my sense of [the] meetings is that there is an atmosphere of suppressed panic about the oil price, and about the danger of a serious crisis".[236]

The following week the oil price went over $50 per barrel for the first time.[237] Within days it had crossed $54 in New York and $51 in London.[238] By the end of the month it had crossed $55. Every small jolt to the market pushed it to new levels. The latest concerns involved tight heating-oil supplies ahead of winter, strong oil demand in China and fears about a planned petroleum-industry lockout in Norway.[239]

In an extraordinary move that would hardly have the effect of calming fears about supply, at the end of October OPEC called on the US to open its 670-million-barrel Strategic Petroleum Reserve to help "cool" the prices. This reserve, kept in caverns in Louisiana salt mines, is supposed to be for major emergencies. No, said the US, there is plenty of oil in OPEC countries. Get pumping.[240]

Lord Browne once more took to the field to try and calm the jitters. "It is not helpful for the world to believe that it is running out of oil," he said. "We are evidently not."[241]

The next day, both BP and Exxon announced yet another set of record profits, and Shell warned that yet another reserves downgrade was in the offing. Shell CEO Jeroen van der Veer broke the usual industry solidarity at this point. "There is something strange going on in this industry," he told the *Economist*. He voiced suspicions that his company was merely the first to face up to a problem that was actually industry-wide. "I would bet money that is the case," one analyst responded.[242] But that didn't help Shell much. The continuing uncertainty over reserves meant that the once seemingly invincible giant had to announce its AGM would be delayed.

ENTERING 2005: THE GREAT GLOBAL ENERGY CRISIS TAKES SHAPE

The year 2005 began with Saudi Aramco deciding to drop the veil a little on its half-century of secrecy in an effort to calm the growing fears. Being a state-owned creature of the biggest oil producer, it is the biggest oil company in the world by far: some twenty times bigger than the biggest private-sector company, ExxonMobil. To counter the growing view that it has only 1 million barrels per day of spare capacity on top of its 9 million barrels a day of current production, the company promised spare capacity could be maintained at 1.5 to 2, and even professed that Aramco's technical experts would be prepared to discuss details of how they will be able to hike production to 15 million barrels a day, and keep it at and above those levels for fifty years to come. But they refused to countenance letting independent auditors in. And still some 90 percent of their oil comes from just seven fields that are an average of forty-seven years old, including 5 million barrels a day from just one, Ghawar.[243] Energy investment banker Matthew Simmons duly told the Arab news agency al-Jazeera that he thinks Saudi oil may already have peaked because they have harmed the structure of the reservoirs in earlier phases of rapid pumping while trying to bring oil prices down.[244]

Entering March, the New York Mercantile Exchange reported its first trade of an oil option with a strike price of $100.[245] Oil prices predictably surged again, this time to a new record of $57, with analysts attributing the new peak to speculators such as hedge funds.[246]

At this stage the International Energy Agency, hitherto a standard bearer in the "no need to worry" brigade, called for an emergency oil-conservation plan. A leaked report concluded that restraint in importing countries "may be attractive in periods of high oil prices to relieve demand pressure". These words went straight on the front page of the *Financial Times*.[247]

In March, ChevronTexaco made a bid for middle-ranking oil company Unocal. Why would they do that if there was a fair chance oil prices might come down again? The buyer would be left looking rather silly, and much out of pocket, if that happened. One commentator in the *San Francisco Chronicle* ventured an opinion. "Reading between the lines, that means only one thing. Peak oil. We're basically there."[248]

April saw another landmark in the unfolding story: the first mainstream bank to say that Saudi Arabian oil has peaked. A Bank of Montreal report concluded "Hubbert's peak has arrived in Saudi Arabia".[249]

25 April 2005, Edinburgh

I sit on a stage facing a packed auditorium with Colin Campbell, Chris Skrebowski and Matthew Simmons. This is the first conference on oil depletion in the oil-rich nation of Scotland. The audience is firing questions at us. A Scottish celebrity broadcaster is chairing. It is late afternoon, and we have all given long talks earlier in the day. The event has been organized by a group of concerned private citizens who usually meet in a pub, yet the attendees include national and international media, politicians, civil servants, oil-industry executives, business people from many other industries and academics.

There is another category of attendee, one that is much on my mind as I think about the coming great global energy crisis. The leaders of the British National Party, otherwise known as the Fascists, have shown up. Five of the people in this sea of faces in front of me lead the nation's card-carrying Nazis. I scan the audience, trying to guess which ones they are. I can well imagine why they are here. Type "Great Depression" and "Rise of Nazis" into Google and the answer would doubtless leap out. If there is mileage in this business of the oil peak, their thinking must go, the markets will topple, recession will turn into depression, the unemployment queues will explode. There will be armies of seriously disaffected people. The fertile breeding grounds of Nazism, last seen in the 1930s, will once again be with us: playgrounds for any sick thug with a working knowledge of the politics of power.

But they are not wearing their brown shirts and swastikas today. They are just sitting there, listening, learning and scheming in the privacy of their heads. I feel – how shall I put it – very ambivalent about this.

Today is the first time I have met Matthew Simmons. He flew in this morning from Houston in his Lear Jet. He is everything I imagined him to be: sharp, clear, passionate about the problem, and bewildered that more people can't see the crisis. I wonder what his peers and employees think of him, interrupting his profitable investment banking to fly half-way around the world to speak at the invitation of a bunch of British pub activists. He talked this morning of the urgent need for an early warning system and greater transparency. Some 95 percent of the world's supposedly proven oil reserves are unaudited, he said, with an affronted air. Water and gas injection into oilfields can artificially prop up high pressures, but when pressures finally ebb there is no secondary game to play. We need to know for sure what is going on, given the stakes. Ten good analysts could find out in under a month where the global peak of production lies with data from the hundred biggest fields, he figures. But most of the data is kept secret by the companies. So we have to guess. And his best guess? The peak is close or even past. But let's find out! Let's deal with it! Why are we asleep at the wheel?

Not unnaturally, many of the questions are now being addressed to him. "When should I sell my BP shares?" asks one man. Simmons shrugs. "The big companies can't drill their way to growth," he replies.

"How likely is it that Saudi oil has already peaked?" asks another. "There is a chance it peaked as long ago as 1981," Simmons responds.

"Was the war in Iraq an oil grab?" asks a third. "No way," he insists.

Lest we forget, Matthew Simmons has been an energy adviser to George Bush, which is one of the many reasons why his story should be being taken far more seriously than it is. The celebrity chairman throws that one back to the audience and asks for a show of hands as to who thinks it was a war for oil. A forest of arms go up – the vast majority. I glance at Simmons, who is sitting next to me. "I'm really shocked," he says. And he looks it.

He proves to be more of an optimist than I had imagined. The world can cope with post-peak oil, he asserts. The aftermath can be managed, if it is truly understood, by using the ultra-high revenue flows from the second half of the oil era in crash programmes to modernize energy infrastructure, maximize energy efficiency, create less energy-intensive transportation, and by directing oil to high-value uses. There will need to be consumption quotas. The industrialized world will need to go on an energy diet, while the likes of China and India get a bigger slice of the cake. Finally, there will need to be R&D Manhattan Projects to invent a new energy plan for the world.

But have we waited too long? Simmons doesn't know. "I hope we have time," he says. "This is, er, a major issue."

And I scan the faces again, Fascist-spotting, wondering if these men I am with on stage and I are right, and actually hoping we are not. I have a recurrent argument with one of my friends. He is 100 percent certain the oil production topping point will happen this decade, and create what he calls the peak panic point with it. I tell him I am 98 percent certain. I am a scientist, I say. You can be 100 percent certain of very little in this world, with my kind of training. There is most certainly some potential for events to turn in a way that I and others have not thought of.

I spot a candidate in the audience. Yes, with a little imagination it is easy to see him wielding a baseball bat somewhere other than a playing field. I think of my one-year-old grandson and the decades ahead of him.

Yes, please let me be wrong.

CHAPTER 8

What can we do about it?

This book has endeavoured, so far, to prove the case for two big arguments that are the starting premises for this final chapter. First, there is plenty of oil and gas left, but not enough to feed growing global energy demand for much longer. The oil topping point, otherwise known as the peak of production, will be reached in the 2006–2010 window and when the market realizes this, severe economic trauma will ensue. Second, global warming is a real, present and fast-growing danger. It will destroy economies and ecosystems if more than a small fraction of remaining coal is burnt. Burning most of the remaining oil and gas will have the same effect, wherever the oil and gas topping points lie.

Beyond these premises, I now make five arguments, one leading on from the other:

1. It will be possible to replace oil, gas and coal completely with a plentiful supply of renewable energy, and faster than most people think. (We can think of this – in a very real sense – as "endless power", and we will get to that promised land one day. When we do, we'll wonder why it ever took us so long.)

2. *However* – a very big however – the shortfall between current

198

expectation of oil supply and actual availability will be such that neither gas, nor renewables, nor liquids from gas and coal, nor nuclear, nor any combination thereof, will be able to plug the gap in time to head off the economic trauma resulting from the oil topping point. (Stated another way, we've left it too late. A lot of people are going to be pretty angry about this, come the panic … and want to know why. Watch out politicians and oil companies.)

3. Renewable energy and fuel use, alongside energy efficiency, will increasingly substitute for oil and gas, growing explosively whatever happens. (And OK, I can hear it already: "You run a renewable-energy company. Talk about vested interests! You would say that wouldn't you?" But the reason I run a renewable-energy company is fear of global warming, primarily, as my track record shows. So I invite critics to judge my case on the arguments.)

4. However – another very big however – amid the ruins of the old energy modus operandi many will try to turn to coal. This means that the extent to which renewable energy grows explosively *instead* of coal expansion, rather than alongside it, will determine whether economies and ecosystems can survive the global warming threat. (Paraphrasing the insurance guru who featured in Chapter 5, we have to think silicon and matters overground, not carbon and matters underground, in order to keep the world habitable for our children and theirs.)

5. There is much that people can do to influence the outcome of this struggle to increase renewables' production faster than coal, hence to ameliorate the worst excesses of the global

energy crisis, and to create a better society in the process. (And this, really, is the most important point of all.)

1. IT WILL BE POSSIBLE TO REPLACE OIL, GAS AND COAL COMPLETELY WITH A PLENTIFUL SUPPLY OF RENEWABLE ENERGY, AND FASTER THAN MOST PEOPLE THINK

Shell employs roomfuls of clever people just to think about the future. They are called scenario planners. In their 2001 book of scenarios, Shell's planners mention that renewable energy holds the potential to power a future world populated with 10 billion people, and do so with ease. The needs of the 10 billion can be met even in the unlikely and undesirable event that all of them use energy at levels well above the average per-capita consumption today in the EU.[250] The Shell futurists mention this almost in passing, in the caption of a diagram showing the continent-by-continent potential for individual renewable-energy technologies to contribute to such a power-rich future. Working for an oil and gas giant as they do, it is perhaps no surprise that they fail to explore a scenario wherein something resembling this renewable power-rich future comes to pass. Let *us* consider it.

In such a discussion, it is important to emphasize at the outset what might be called the "big portfolio" approach to the retreat from fossil fuels. No one technology would be needed to produce all of a nation's energy demand, or even come close. We can mix and match among the family of renewable-energy technologies, because it is a family with a lot of brothers and sisters. We'll have a look at the family in a moment, but we will also need to consider two vital additional components of the story: renewable-energy storage and energy efficiency. As fossil-fuel diehards are fond of saying, there is not much point in having energy from the elements if it can't be stored. You must have heard them. "What happens when the wind doesn't blow?", ha ha. "What happens when the sun doesn't shine?", chortle chortle. This is where fuel cells, hydrogen and batteries enter the equation. Similarly, if the world's routine energy-demand profligacy can be deconstructed with efficiency measures, the mountain to climb without oil, gas and coal can be made a lot less steep. Let me put that more strongly: with simple, economic, smart, short-payback investments, we could blow the top half of the mountain off, much in the noble tradition of Mount St Helens. In what follows I am going to consider renewables, storage and efficiency separately, though there will be profound overlaps between the three when the world is forced by the oil topping point to attack its energy-supply problems seriously.

Renewables: a big family of options

The first thing you notice about the Shell diagram showing renewable supply meeting the energy demands of 10 billion people wasting energy at the level your average wasteful European does today is that solar power is the biggest potential contributor to such a hypothetical future. At one level, this is not perhaps surprising. Enough light falls on the surface of the planet each day to power human society many thousands of times over. Using solar photovoltaic (PV) cells, the world's current energy demand – all forms of energy use including transport – could be met using a tiny fraction of the planet's land surface.[251] For example, the total electricity-generating capacity of all the existing power stations in the world today, of all types, could be created by covering an area of the Sahara desert some 600 square kilometres with solar PV.[252] Even in the cloudy UK, more electricity than the nation currently uses could be generated by putting PV roof tiles on all suitable roofs.[253] Solar thermal technology, which can be used for both heating and electricity production, holds no less potential. This member of the family uses collectors to absorb heat from sunlight and then to heat liquid. The heat can then be released from the liquid into a storage tank for use in a building. Where this technique is used for electricity generation, devices that concentrate sunlight, such as mirrored curved reflectors, heat liquid to very high temperatures, creating steam that drives turbines. "Solar farm" power plants of both solar PV and solar thermal collectors exist today only in small numbers, but will be a common sight – especially on otherwise useless scrub land in the sunbelt – once the solar revolution takes off.

Wind power also plays a huge role. America could provide all the electricity it uses today from the wind-power potential of just three states: Texas, North Dakota and Kansas.[254] Europe's electricity demand could be met using offshore wind farms alone.[255] In the UK's case, only a tiny fraction of suitable offshore areas would be needed to meet the nation's total current demand. The potential of solar and wind extend beyond electricity and heating to transportation. Sufficient hydrogen to fuel every highway vehicle in the United States could be generated, for example, with the wind potential of two states alone, the Dakotas.[256]

As well as solar and wind, our renewable-energy options include marine energy, hydropower, biomass, combined heat and power, heat pumps and biofuels. Let me quickly summarize these technologies, and their potential, in turn, before moving on to the close relatives of the family, storage and efficiency.

Both tides and waves can be used to generate marine power. To tap the energy of tides, gates and turbines are installed along a dam or barrage across an estuary or bay. When the height of the water builds up to the right level on either side of the barrage – on both incoming and outgoing tides – the gates are opened, and the water flows through the turbines, turning electrical generators to make electricity. A 240-megawatt station has been operating in a small estuary in Brittany, France, for forty years, but it is the only example in the whole of Europe.[257] In the UK, for example, if all exploitable estuaries were utilized in this way, some 15 percent of national electricity demand could be met, according to government estimates.[258]

To tap the energy of waves, a variety of devices have been designed, deploying turbines either on the shore, near-shore, or offshore in open water. Waves effectively concentrate the energy of wind: because water is much denser than air, the energy needed to move a certain volume of water is much greater than that needed to move the same volume of air. For this reason, a small turbine past or through which a wave passes can generate the same power as a much bigger wind turbine. A promising technology called Pelamis, under test offshore Scotland, consists of articulated cylindrical sections that are hinged so that they can move with the waves, powering as they do so hydraulic motors that generate electricity. A prototype 120 metres long and 3.5 metres wide generates 750 kilowatts. A "wave farm" of such devices spanning just a square kilometre of ocean would generate enough electricity for tens of thousands of homes.[259] We are not space constrained when it comes to waves.

Similarly, currents in rivers can be tapped by small run-of-river turbines. Such technology is called micro-hydropower if the turbines are less than 1.25 megawatts. There is huge potential here. It is quite amazing how much electricity can be generated by a little moving water. In the UK for example, former mill sites alone have a combined generating capacity of one and maybe two typical nuclear or coal plants.[260]

Biomass is an important renewable resource because of both its potential scale and the fact that it can be used in a way that produces no net greenhouse-gas emissions. Biomass fuels are of three types: waste by-products (from agriculture, forestry and the urban environment), energy crops and processed fuels (for example wood pellets made from sawdust). So long as the plant matter used is replaced by regrowth – hardly a problem when it comes to agriculture – there is no net build-up of greenhouse gases in the atmosphere. Heat can be generated from these fuels by straightforward combustion in power plants and boilers, or by a group of higher-efficiency processes: anaerobic digestion, gasification and pyrolysis.[261] The heat from burning biomass can be used directly, or used in a turbine to generate electricity. Again, the potential is vast. Take, for example, production of straw in the UK. Suppose one third of the annual production was burnt in biomass power plants: some 8 million tonnes of biomass material. This one crop residue would generate 3 percent of UK electricity. Instead, most such residues simply go to waste. As for energy crops such as fast-growing willow and Miscanthus, there is strong potential. In round figures, if 10 percent of the UK's 20 million hectares of agricultural land were taken up with short rotation coppice – some 2 million hectares – biomass could meet 10 percent of the UK's current electricity demand. Note that 0.6 million hectares of UK agricultural land is set aside as things stand, i.e. essentially unused.[262]

Combined heat and power (CHP), as the name implies, allows any combustible fuel to be used with high efficiency. Generating electricity alone in power plants is typically only 30–40 percent efficient, because so much heat is lost. CHP generators, which use the heat produced as well as the electricity, work at around 80 percent efficiency, generating around three times more heat energy than electricity.[263] All biomass power plants could also be CHP plants. We don't have to be purist though. Biomass can and is burnt along with fossil fuels in co-generation plants, with consequent reduction in fossil-fuel use and greenhouse-gas emissions.

This completes the renewable toolkit for power plants, large and small, in or distant from the buildings they are heating or lighting. The renewable technologies that can be used within buildings are often referred to by the catch-all term "micropower". The renewable micropower family consists of both types of solar, micro-wind (small wind turbines on roofs), biomass boilers, micro-CHP (if the fuel is biomass) and ground-sourced heat pumps. The role they play in electricity generation can be thought of as "embedded generation", in that they are embedded in the national electricity grid. When the renewable micropower technologies are in common use, that grid will be a very different animal from the one that exists now. For the most part electrons flow in the grid one way only today: from the giant and dirty power plants we have tended to favour historically to the end user in their home, office, factory or whatever. This requires high voltages, transmission lines running long distances, and transformer stations to step the voltage down for use in buildings at the end of the line. With embedded generation, the distances travelled by the electrons is shorter, and everything is manageable with lower voltages. Many electrons are used right there in the building where they are generated. Smaller grids can be used. The whole infrastructure tends to resemble the internet, where many "distributed" computers are used in preference to the giant centralized mainframes of old. More and more people are starting to refer to the "energy internet", in fact, when they talk about all this.[264] Small grids can indeed be isolated from the national grid. A new town, village or community might have its own "private wire" network. Some do today, although this is still rare.

To give just one illustration of one of the renewable micropower family at work today, ground-sourced heat pumps are already in common use across the rural United States. These pumps extract the solar energy stored at a constant 11–12°C a few metres down in the ground. Working like a reverse refrigerator, using a mix of chilled water and antifreeze in a coil of pipe, they pump this heat into the home for hot water and heating. They can be used in reverse, in hot summers, to pump heat out of the building into the ground, thereby obviating air conditioning. Some electricity is needed to work the pump, but of course that can come from other renewable sources.

It is all very well heating and electrifying buildings, but how are we going to get around in this brave new world? Once again, we are not stumped for options, only the imagination to see them and the will to make the changes. Automotive fuel for direct use can be made in abundance from plant matter. Biodiesel can be made from such plant matter as soyabean, vegetable or rapeseed oil and used directly in cars with diesel engines. In 2004, US manufacturers like DaimlerChrysler and General Motors suddenly began to take a serious interest in biodiesel. Volkswagen has said it will use biodiesel as the best way to compete with Toyota's immensely popular fuel-efficient hybrid car, the Prius. DaimlerChrysler has said it will fill all new Jeep Liberty vehicles with biodiesel.[265] Such interest can be explained at least in part by targets set by governments. Europe, for example, has a target for biofuels made from all agricultural forestry and organic waste: 2 percent of fuel use by 2005 and 10 percent by 2010. Beyond biodiesel, there is ethanol – made from maize, and being produced in the US today in seventy-five small subsidised refineries (with a further twelve under construction). On top of this, methane gas can be made, in principle in vast quantities, from anaerobic digestion of any organic waste, plant or animal, compressed and used as a fuel.

Plant matter can also be used rather than oil to make plastics and other chemicals important in energy technology. At the 2003 Detroit motor show, Ford introduced a concept called the Model U green car with engine oil made from sunflower seeds and seats made from soyabeans. The company was merely extending a tradition: its founder, Henry Ford, had used soyabeans rather than oil to make plastic, and produced a car completely made from plants, as long ago as 1941. Such cars may never go on sale as things stand, but they could, given the right conditions. Their engines could be filled with fuel made from plants. Biofuel refineries could be built faster, targets could be made more ambitious, incentives of many kinds could be deployed to turn the Fords of this world from tinkerers with biofuelled biovehicles to mass producers.[266]

Then there is hydrogen, in which the auto giants are showing more than a passing interest.

Storage: the road to the hydrogen economy

Hydrogen is not a fuel in the strict sense, but rather an energy-storage medium, in that it is not found in nature unless combined with other elements. It is usable in fuel cells or as a solid or liquid fuel. A fuel cell is a modular device that chemically recombines hydrogen with oxygen on a catalytic membrane producing an electric current as it does so, plus a single and uniquely unproblematic waste product: pure hot water. There are several different types of fuel cell, but they all work in this basic way.[267] The oxygen can typically come from the air, and the hydrogen can be made from any other energy source. At the undesirable end of the spectrum, because a lot of carbon dioxide is emitted along the way, coal can be reformed. (A simple way to think of reforming is taking the hydrogen out of the hydrogen-and-carbon mix in hydrocarbons.)[268] At the desirable end of the spectrum, because no greenhouse-gas emissions result, renewable energy can be used for electrolysis to generate the hydrogen. (A simple way to think of this process is taking the H_2 out of H_2O.)[269]

Fuel cells use their hydrogen about twice as efficiently as internal combustion engines use their gasoline, per unit of fuel. Currently, however, they cost one hundred times more per unit of power, and hydrogen itself is about five times as expensive as gasoline.[270] General Motors, DaimlerChrysler and Shell have all already invested more than a billion dollars each in fuel-cell research and development (R&D), trying to close this gap, and many other corporations have significant R&D programmes under way. Governments are encouraging their national industries to view the search for commercial hydrogen vehicles as a race. General Motors, endeavouring to set the pace for the US government, which has made a $1.2 billion R&D commitment over five years, has said publicly that it expects to begin selling hydrogen-fuel-cell vehicles in 2010. It may have to hurry, because the Chinese government is aiming to be the world leader in hydrogen-fuel-cell-powered cars, and has been supporting R&D at the level of $200 million per year in the past few years. Local industries have so far produced more than a thousand patent applications in the area of fuel-cell technology.[271] China is already believed to be the number two producer in the 50 million tonnes of hydrogen produced per year globally, and Japan has a specific goal of 50,000 fuel-cell vehicles on the road by 2010. One hundred and seventy-two prototype hydrogen cars and eighty-seven hydrogen filling stations have already been created worldwide.[272]

Fuel-cell cars and buses are to be found today on the roads of many cities around the world, but not yet at prices affordable by most people. The key to making this happen – the central thrust of the international R&D race – is reducing their size and weight. In buildings that is less of a problem, and bulkier fuel cells are already to be found in pioneering green buildings, generating both electricity and useful hot water.[273]

Others have their doubts about fuel cells being the best way to use hydrogen. BMW, for example, has opted for use of the gas in liquid or solid form – enabled by the storage under pressure of solids called hydrides – and hydrogen gas-filling stations have already been built on an experimental basis in Germany. BMW, in fact, began a massive marketing push in April 2001 to reposition itself as the lead pioneer of the hydrogen age. The top brass of the company toured major world cities, including London, where their event was staged at the Science Museum. I have never been to such an extravagant reception, before or since. Beautiful blonde women served champagne and canapés among hydrogen paraphernalia and glossy posters, with a gleaming BMW on centre stage, run on liquid-hydrogen fuel and looking just as sexy as its internal-combustion-engine predecessors. Dr Helmut Panke, Chairman of the Board of Management, gave a rousing speech. "The hydrogen age has begun," he announced to a sea of several hundred suits invited from across industry, government and media. "It has to start in our minds." One of the four most senior ministers in the British government spoke. A BBC newsreader moderated.

Events like this make clear an important distinction between the auto and oil industries: the former is not necessarily going to lose its core product when the wheel of change really begins to turn. In that, it is much less threatened than the oil industry. The oil industry will still be in business in the second half of the oil age, of course. It will be providing much-needed hydrocarbons for the chemical industry, for one thing. But it will be a shadow of its former self and, unless the companies take the lead in the renewable revolution, they will no longer be among the largest corporations in the world.

The challenge will be getting the auto industry to the point of no return, where they commit willingly – or are forced to commit – to the mass retooling of their factories for means of propulsion other than the internal combustion engine. The scale of this challenge is nowhere more clear than in the case of battery storage. I drove the General Motors EV1 battery vehicle as long ago as the G8 summit in Denver in the summer of 1999, where the auto giant was making the futuristic-looking electric saloon available for delegates to take a quick spin. Sitting in the spacious driving seat looking at the digital displays had more the feel of being in an aircraft cockpit than a car. Soundlessly the vehicle started, and as I squeezed the accelerator it burst down the wide Denver streets with the power of anything I have ever driven powered by gasoline. Its nickel-hydride batteries were capable then of 160 miles between charges.[274] How often would anyone ever want to go further? Why couldn't General Motors simply mass-produce this car, one might wonder, so bringing the price down to an affordable level? But when the state of California introduced a draft legislative mandate for a small percentage of the cars sold in the state to be zero emissions, trying to encourage the proliferation of EVs in the interests of air quality, General Motors lobbied fiercely against the measure, along with the other auto manufacturers, and forced it to be watered down.

That was then, however. The success of the Toyota Prius in the last few years is forcing a major re-evaluation of attitudes to disruptive technologies among the major manufacturers. The Prius uses a combination of a battery at low speeds and an efficient internal combustion engine at high speeds to achieve sixty miles per gallon of gasoline, compared to the record-low average of the 200 million vehicles in the American fleet last year running at 20.4 miles per gallon. The popularity of the vehicle caught Toyota, by far the world's most successful auto manufacturer, by surprise. The first model, introduced in 2000, sold only 15,000. As of September 2004, Toyota was selling that many a month of the roomier, more powerful second model, with 22,000 customers remaining on waiting lists despite three ramp-ups of production. Ford launched its first hybrid SUV in September 2004, and by 2007, according to industry estimates, there will be some twenty-two hybrid versions of popular models. As General Motors' Vice Chairman Bob Lutz told *Newsweek*: "We can't just sit there as a major corporation and say, 'Trust us, you'll get a fuel cell from us and in the meantime we're not doing anything.' With more and more of our competitors playing the hybrid card, there is just no way we can ignore that." [275]

Promising stuff, and it begs the question of just how much energy efficiency can reduce the target for renewables in replacing fossil fuels, both in the transport sector and in the built environment.

Efficiency: reducing the demand mountain to a hill

Arguably the biggest expert in the world on this subject is Amory Lovins, the Director of the Rocky Mountain Institute in Colorado. In my years as an environmental campaigner, I would often find myself sharing a platform with this sober physicist as he recited, deadpan, and at machine-gun velocity, the killer statistics of energy-efficiency savings, and how many easy-payback investments there were, long term and short term, available in the built environment, on the roads and in the airways. Each time, I would find myself thinking at the end: "Why don't people just *do* this? They don't have to believe in the threat of global warming, or air pollution, if they don't want to. They could just go off and make money." Lovins himself regularly told an anecdote to make the point. A little girl walks down the street and sees a hundred-dollar bill lying on the sidewalk. She says to her wise old grandad: "There's a hundred-dollar bill on the ground!" But he says: "No sweetie, if that was a hundred-dollar bill, someone would have picked it up by now."

In 2004, Lovins and his team published the latest in their exhaustive studies of the efficiency field, this time part-financed by the Pentagon if you please, and very much rooted in the imperatives of our time. He called it *Winning the Oil Endgame: Innovation for Profit, Jobs and Security.*[276] Amazingly, but unsurprisingly to those who have exposed themselves to a few details about energy efficiency, it concludes as follows: "... it will cost *less* to displace all the oil the United States now uses than it will cost to *buy* that oil." To replace oil use with cheaper alternatives in this way, the US would have to invest $180 billion over the next decade, for which the return would be $130 billion in *annual* savings by 2025. To win this jackpot, the investment would need to be made according to four technology strategies pursued step by step: first, using oil twice as efficiently as is the case today; second, substituting biofuels; third, saving natural gas; and fourth, introducing hydrogen. As will be clear from what we have considered above, these are all trends that are under way already.

Doubling the efficiency with which the US uses oil can be achieved with a variety of techniques, especially ultralight-vehicle design, Lovins and his colleagues argue. Good as the efficiency of the current hybrid-electric cars is relative to the gas-guzzling norm, advanced composite or lightweight steel materials can nearly double it at an extra cost re-coupable from fuel savings in about three years. Lovins introduced the concept of such "hypercars" with another team, this time including vehicle designers, in 2000. Beyond the uptake of such smart technology, creative business models and public policies can easily make up the rest of the oil savings. A good example is the introduction of feebates: fees for inefficient vehicles or buildings that are recycled in a revenue-neutral way as rebates for efficient ones.

The vehicle improvements and other savings don't need to break any records. They needn't even be as fast as those actually achieved in America after the 1979 oil shock. The investment required to achieve this goal would be $70 billion of the total $180 billion.

Beyond efficiency savings of one half of all projected oil use, a further quarter can be saved by creating a major domestic US biofuels industry. Lovins and his team figure that rural America can be much strengthened by an assault by biofuels on hydrocarbon markets. Farm income could be boosted by tens of billions of dollars a year and more than 750,000 new jobs created. This would require some $40 billion of the $180 billion total investment.

The "low-hanging fruit" of efficiency savings in natural-gas use save at least half the gas demand projected by the US government in 2025. The saved gas can be in part substituted for oil in a further demand reduction, or it can be recombined to make hydrogen, displacing almost all the rest of US oil. The rest can easily be replaced by renewables.

How inspiring, and yet how conservative it all is. Lovins makes no radical assumptions about explosive growth of renewables or hydrogen. Indeed, he classifies the development of hydrogen as *optional*. If you merely want to cut US oil consumption to the point that no imports are needed, you do the efficiency and biofuel parts. If you do the hydrogen and renewables part as well, you can wean the country off oil totally.

Some of the implications are mind boggling, especially the $180 billion investment needed to hit the plan in the context of the current high oil prices, not to mention the $2.4 trillion investment needed to meet projected oil demand over the next decade according to Goldman Sachs, or the record cash hoards of the oil companies.[277] Exxon alone has amassed $25 billion at the time of writing (in the face of investor pressure to shell some of it out as dividends, or go for a mega-acquisition).[278] "The United States' economy already pays that much," Lovins observes of the $180 billion, "with zero return, every time the oil price spikes up as it has done in 2004."

Not to mention what will be paid for oil after the global production topping point.

How quickly can we withdraw from oil, gas and coal?

We can think of the renewable technologies as solar technologies in the broad sense: most are driven either directly or indirectly by sunlight falling on Earth. The sun creates differential warming in the atmosphere that creates winds, and so drives wind power. Winds create waves at sea, which drive wave-power devices. The sun creates differential warming in the ocean, which creates currents, which can drive marine turbines. Sunlight is needed for photosynthesis, and hence is at the root of plants useable as fuel. The only type of renewable energy unrelated to sunlight in some way is tidal power, which is made possible by the gravitational pull of the moon. We can think of the whole process of exploiting and commercializing these technologies as "solarization". This technologies themselves are renewable in the sense that they hold the potential for generating power as long as light falls on the planet: effectively, endless power.

It is far better to use all members of the solar-technologies family than to concentrate on one or two, obviously. The two main reasons for this are security in diversity and the ability to meet variable energy loads with spare capacity by mixing the means of supply. For example, solar can meet much of the peak air-conditioning load on summer days while biomass can contribute best to winter heating; marine can provide baseload while wind chips in where it can; and so on.

One of Shell's most famous scientists saw all this, a long time ago. "Our culture doesn't know how to deal with a levelling off or a decline but it will have to," said M. King Hubbert. "We have to steer ourselves into a stable state with as little catastrophe as possible. We should be looking for other sources of energy. There's only one big enough. It's free, and it's good for at least a billion years. That's the sun." [279]

When I began my time as an environmentalist, in 1989, the protestations my colleagues and I made that renewable energy could displace fossil fuels and run the world were ridiculed by energy experts and officialdom as naive wishful thinking. Now, more than a decade later, such views can be found in the heart of government, at least in Europe. The UK government published a report in 2003 that concluded as follows: "… it would be technologically and economically feasible to move to a low carbon-emissions path, and achieve a virtually zero carbon-energy system in the long term, if we used energy more efficiently and developed and used low-carbon technologies." [280] Among the low-carbon technologies on offer, the government report places heavy emphasis on renewable energy and hydrogen, rather than nuclear power. Of solar energy, the report concludes: "[It] alone could meet world energy demand by using less than 1 percent of land currently used for agriculture." Prime Minister Tony Blair used these same words in the speech he gave launching the UK Energy White Paper. I sat there watching him do it, ten feet away in the front row. I was momentarily tempted to leap to my feet and shout "So why don't you invest in it like the Germans and Japanese then?"

Microcosms of what could be done on a much bigger, even international, scale can be found already on the local government scene. Take the small Surrey town of Woking in the UK. Its borough council has cut carbon-dioxide emissions by fully 77 percent – yes, more than three-quarters – since 1990 using a hybrid-energy system involving private wires, CHP (mostly natural gas but some biomass), solar PV, and energy efficiency, plus some fuel cells and absorption chillers.[281] I toured this inspirational scene with Woking's Energy Manager, Allan Jones. Proudly he showed me old people's homes and housing estates that had been made into their own mini energy worlds. The UK grid could go down for ever, and these folks would have their own heating and electricity year-round. The technologies work in perfect harmony. The CHP units generate heating when needed in winter, and lots of electricity along with it when the PV is not working at its best. The PV generates plenty of electricity in the summer, when the heating isn't needed, meaning the CHP can't generate much electricity. The Greater London Authority has now hired Allan Jones to do the same for London. OK fella, the Mayor said, you did this for a town of 80,000. Now do it for my city of 8 million.

When summoning the evidence to tackle the question of how quickly we could solarize, success stories like those in Woking provide real inspiration. So too do powerful out-of-the-box ideas. Back to Amory Lovins for a moment for one of those. Suppose in Woking all the cars had fuel cells. Suppose that, when they were parked, they were plugged into the electricity grid, either the national grid or one of the private wire grids, trickling electricity into the wires. How much do you think that would add to the picture? Lovins and his colleagues argue that there are two keys to unlocking hydrogen's potential. One is early deployment of super-efficient vehicles, obviously, which shrink the fuel cells so that they're affordable, and make the hydrogen fuel tanks packageable into the vehicle body. The other is the added value of using the hydrogen fuel cells while the vehicle is parked, rather than just letting them sit there uselessly. Each of these key developments would accelerate the other by building volume and so cutting cost. But beyond the boost hydrogen technology would be given, think of the scale of the energy made available. Currently, some 200 million of the world's 700 million cars and trucks travel and park around America. If they were all hydrogen-fuel-cell vehicles they would have *many* times the grid's total generating capacity. What need then for nuclear, coal and gas-fired power plants?[282]

Faced with all the evidence of potential in the wider renewable/storage/efficiency family, inspirational examples like Woking, and amazing but feasible ideas like plug-in fuel-cell vehicles, what can we now say about how quickly the world could solarize if it really wanted to?

Primary fuel and other technology substitutions usually take fifty years, Amory Lovins and his colleagues observe, somewhat discouragingly. They are thinking here of wood to coal to oil to gas, and the growth of railways and electrification. Hydrogen can penetrate more quickly, they argue, because of the scope for innovation at small scale, taking maybe taking twenty-five or thirty years.

Are there any circumstances in which things could move faster than this? Surely. Consider the automobile. Horses were the dominant means of transportation in 1900. The early automobiles, the horseless carriages, were laughed at. They were expensive, slow, unreliable, and had few suitable roads to travel on. Then Henry Ford started mass-producing them, bringing down costs sharply. In parallel, he lobbied for better roads. In 1900, eight thousand autos were registered in the US. By 1912, there were over nine hundred thousand. Within just a decade automobiles had gone from nowhere to almost everywhere.

Of course, that was with a much lower population and a much less pervasive infrastructure. But remember Sheik Yamani's warning to his fellow oil ministers in 1981 that Western countries would be able, if forced by high oil prices, to find alternative sources of energy within ten years? Surely, with twenty years of further research under our belts now, that's easily achievable? A post-9/11 reluctance to depend on overseas sources of fuel is indeed focusing intense interest on domestic renewable energy potential in many countries. New renewable and power-storage technologies are being developed at an accelerating pace. Big companies like Sharp, Sanyo, RWE and General Electric have all recently begun investing on a scale hitherto unknown in the embryonic solarization industries. Social and personal priorities are beginning to focus on energy. Socially screened investment funds that do not invest in oil companies but will invest in renewables are proliferating as citizen investors seek an outlet for their concerns about the environment. More than one dollar in every ten in the US now goes into such funds.[283] In parallel, fear of power cuts such as those experienced in the wake of the 2003 California energy crisis is driving citizen and government interest in renewable power. Governments national and local are looking at their often fragile electricity infrastructures and wondering if the same might happen to them. The people of San Francisco recently voted in a $100 million bond for solar and wind power in their city with the promise of more to come. Other cities in other American states are considering the same kind of step. In the UK, forty local governments are telling developers they risk not winning planning permission for new buildings unless they have 10 percent energy generation onsite with renewables. Markets in renewable technology, though still small, are already growing at some of the fastest rates to be found. Serious money is waking up to this. Renewable energy involves, after all, an entire family of classically "disruptive" technologies – in the same way that the internal combustion engine disrupted the horse market, or the microcomputer the mainframe market – poised to invade a $4 trillion market: global energy supply *plus* construction.[284] Almost nowhere else can an investor hope to find such growth potential in the modern world.

Amory Lovins and his colleagues operate in a world where credibility is all, and where thinking *too* far outside the norms is a recipe for a serious downgrading of stock in a fragile peer group. Hence the conservative assumptions built into their oil endgame study, visionary as it otherwise is in so many ways. Why the heck should hydrogen be *optional?* I ask. Why such a small role for renewables in the story? It falls to others more careless of their reputations, and that includes me, to reach out for the heights of optimism. Suppose all the market drivers align, I ask myself. Suppose we live through a few years of exploding interest in renewable energy even before the oil production topping point hits. Suppose we are lucky, and the topping point happens at the end of the decade rather than this year or next. How fast then could oil and gas then be supplanted? On my optimistic days, I think the process of solarization could happen with the same speed as the oil-auto revolution. Ten years, say, at a low level of probability.[285]

Would that be fast enough to head off the economic trauma that an early oil topping point entails?

2. THE SHORTFALL BETWEEN CURRENT EXPECTATION OF OIL SUPPLY AND ACTUAL AVAILABILITY WILL BE SUCH THAT NEITHER GAS, NOR RENEWABLES, NOR LIQUIDS FROM GAS AND COAL, NOR NUCLEAR – NOR ANY COMBINATION THEREOF – WILL BE ABLE TO PLUG THE GAP IN TIME TO HEAD OFF ECONOMIC TRAUMA AS A RESULT OF THE OIL TOPPING POINT

Lead times

If the oil topping point happens this decade, even accepting the optimistic thoughts above, we are in trouble. "There really aren't any good energy solutions for bridges, to buy some time, from oil and gas to the alternatives," says Matthew Simmons. "The only alternative right now is to shrink our economies."[286] As John McGaughey of *World Energy Review* puts it, "… anything that might be done to mitigate an oil reserves problem, such as oil shale or coal liquefaction or the hydrogen economy, is going to take twenty years or more to come on line."[287] Conservation would take a decade or more.

It is easy to see why such views seem reasonable. The realization that growing supplies of cheap oil are no longer available will dawn on the energy traders at some point rather soon, and as we have seen, reductions in global oil supply of only a few percent in the past have been enough to trigger panic. The first thing governments, industry and populations will do is look for energy-saving economies. Low-hanging fruit, such as car-pooling programmes and bans on Sunday driving, may buy some time. But with the early toppers projecting a 2 percent depletion per year against widely expected oil demand *increases* of 2 percent and more, alternative supplies will very soon become imperative. And here they won't find big enough markets to plug the gaps, as things stand, or will conceivably be standing, by the end of the decade.

Never mind economics for the moment, let us just think about the timing. The time it takes to build liquids-from-gas plants, liquids-from-coal plants, biofuel plants, hydrogen fuel plants, and hydrogen fuel-cell factories (before assembly in autos) is measured in years rather than months.[288] OK, renewable micropower plants can be installed in as little as an afternoon, as solar installers have demonstrated hundreds of times over on roofs around the UK. But there is a problem here too: demand for solar PV and the other types of micropower kit has to met at the factory gate, and factories don't go up in an afternoon. It might take as long as 18 months to build a giant solar PV manufacturing plant from scratch, even with the stops pulled out.

Nuclear inadequacies

In recent years the nuclear industry and its supporters in government have sought to position their technology for a revival after years in the doldrums. George W. Bush's first-term energy policy proposed a new nuclear programme, and in Tony Blair's third term a pro-nuclear push is under way in the wake of the UK's 2005 election. Given the importance of American and British nuclear companies to the global nuclear industry, ardent nuclear lobbies in other countries watch these efforts hopefully.

I do not think the revivalists will succeed for five main reasons: timing, investment, terrorism, waste and track record.

Timing

The timing problem is even worse with nuclear than it is with liquids from gas, liquids from coal, biofuels, hydrogen and fuel cells. In the UK, nuclear expert Gordon MacKerron professes that "there is no realistic chance, given current politics, that nuclear power could deliver new power before about 2020".[289] He draws this conclusion because, in a country where no new reactor building has been agreed for fifteen years, no government could feasibly start a major programme of construction without a period of public consultation. Indeed, the UK government's 2003 Energy White Paper, which rejected nuclear power at the time, specified that any resumption of reactor building could not happen without extensive consultation.[290] Given that no nuclear build has been sanctioned in the vast majority of industrialized countries for many years, this will be true of many nations. A typical UK public consultation would take until 2008 or 2009, MacKerron argues. Siting, licensing and local public-inquiry processes would take until around 2013. Only then could construction begin. Historically, the Japanese have rushed through reactor-build programmes in five years, but more usually it has taken ten or more from planning to first power generation. By that time you would only be replacing old nuclear plants, many of which are at or beyond their planned lifetimes even today. In terms of overall national energy supply, this would require a major reactor-building programme just to stand still, as it were.

Just think what renewable energy and energy-efficiency markets could be doing by 2020, given even a fraction of the governmental and institutional support nuclear has been given for the last half-century. Come to think of it, just imagine what would they could be doing *tomorrow*. When I moved into a solar-powered house a few years ago, I cut the electricity demand of the dwelling by more than two-thirds virtually overnight simply by replacing the lights and appliances with the most energy-efficient models available. That was before the PV roof cancelled demand for other electricity supply out completely. It wouldn't take an unimaginable number of such buildings to wipe out the current 23 percent nuclear market share in British electricity generation.

Investment

Most of the world's energy markets are liberalizing, and in liberalizing markets decisions about what electricity generating plant gets built and what doesn't are ultimately made by investors, not governments. As things stand – simply and graphically stated – financial institutions without exception ignore nuclear power as a repository for their investment dollars because of nuclear economics. Many opponents of nuclear begin and end their case with this argument. Financiers simply won't prove persuadable that nuclear should be financed, they say, and it easy to see why. First, the long planning and construction times for nuclear mean a wait of at least seven years to see returns on capital invested, whereas with combined-cycle gas turbines (CCGTs) and large-scale wind, planning and building can happen in the order of a couple of years. Second, the costs of CCGTs are a known entity in the marketplace. Nuclear costs, with so many open liabilities from unsolved waste problems and potential accidents stretching out into the payback period, manifestly are not. Third, total generating costs for nuclear are very sensitive to miscalculations with performance guarantees, given that 70 percent of the total generating cost involves up-front capital as opposed to capital for fuel and running costs. The proposed next generation of reactors have no proven track record, and the nuclear industry has a long history of over-promising and under-delivering in such matters. CCGTs, meanwhile, come with performance guarantees backed by much operational experience, most of it to investors' liking. Fourth, the nuclear industry needs economies of scale to make the proposed investment work. In the UK, for example, BNFL (British Nuclear Fuels Limited) argues that it needs ten giant reactors (ten 1-gigawatt stations) to bring their costs down to competitive levels: one or two simply won't do it.[291] Here again investors are put off by the spectre of untried technology. Sometimes, when I am finding the going hard on the solar-energy frontier, I cheer myself up by imagining nuclear-industry executives trying to persuade financiers to put multiple billions into the bankrolling of ten nuclear power plants that *might* generate electricity cost-competitive with *today's* best CCGT and wind prices ... around fifteen years from now.

Terrorism

If you accept that a new generation of nuclear plants is needed, you have to accept that a new generation of terrorists is going to find a way to blow up a city or two. This is because wide international proliferation of nuclear-power technology and know-how will be inevitable if the US and UK "lead by example" and introduce a new generation of reactors. Other countries will surely follow that lead. Conscience-stricken nuclear-weapons designer Theodore Taylor, who died in 2004, opposed nuclear power as well as nuclear weapons for this reason. Competent graduate students could build a bomb easily, he used to say, given the wherewithal. With massive civil nuclear-build programmes under way around the world, they would have it.[292]

On top of this, there are the maniacs with degrees in engineering who want to fly fuel-laden jumbo jets into buildings. You wouldn't want such an incident to test a reactor containment vessel, much less the fragile high-level waste tanks at Sellafield. With the wind in the right direction and enough fuel in a jumbo jet, maniac terrorist aviators could kill millions of people.[293]

Waste

When the advocates of civil nuclear power established their industry after the Second World War, few of them thought there would be any problem with "closing the nuclear cycle" – meaning the location of a safe place or places in which to dispose of nuclear waste. Today, no country has established an operating high-level waste repository, and many countries have problems disposing of even low-level waste.

My first experience of the nuclear industry's efforts to paper over this considerable crack happened shortly after I jumped ship to join Greenpeace. NIREX, the agency set up at the time to find a site or sites for nuclear waste, took out press adverts boasting how exhaustive their efforts were. One ad showed a geological cross-section through a candidate site, the Sellafield plant in Cumbria: itself of course a nuclear power site. "The depths we will go to in proving nuclear waste disposal is safe", read their adverts. The picture in the ad showed gently undulating, unfaulted rock strata, with a vertical shaft leading to a deep waste repository. As a geologist, I knew all about the rocks under Sellafield. They were shot through with faults, which displaced the strata considerably, literally chopping them up. Who knew what fluids flowed in those faults, and how the geological forces at work would affect a deep burial chamber for nuclear materials with half-lives measured in many thousands of years? Certainly not NIREX.

Nothing I have seen since has changed my view that geology will be squeezed by politics to make the case for safe disposal. Which perfectly introduces the next argument.

Track record

It is now more than thirty years since a new order for a nuclear plant has been placed in America without being cancelled. One built on Long Island was ready to be turned on, but local opposition killed it in 1994. Similarly in Britain and Japan, the rivers of public opposition run deep and wide. One of the main reasons for this is the nuclear industry's long history of secrecy at best and lying at worst. It would take me a lot of space to chronicle the casebook of cover-ups over the years: cracks in reactors, illegal discharges of radioactive materials, falsified shipment documents, disinformation-dealing about competing technologies, not least renewables. So I am going to rest my case primarily on the arguments about timing, investment, terrorism and waste. Suffice it to say that the nuclear industry's plea for a revival of their technology has been built around three main arguments – nuclear's low greenhouse emissions, the promise of a new generation of modular reactors and their version of economics – and it is worth a quick look at each of these in the context of the industry's track record. My aim here is to provide pointers for those readers who might find the nuclear industry's arguments seductive even after considering the main case above.

Maybe, to save time, I can concede that nuclear power is a low greenhouse-emissions technology. Others wouldn't, and several paragraphs of information about the carbon intensity of uranium-ore mining and milling, steel and concrete in power-station construction, and transportation of fuel and waste, would follow. This to me is off the main point. Even if nuclear power is truly a low-carbon technology, there is no point in having it if you don't need it in the first place because far more attractive options are available. Neither is there any point if you poison the world when you finally get to build your nuclear nirvana. To make any difference in reducing greenhouse-gas emissions, so many reactors would have to be built that the industry would be left with thousands of tonnes of plutonium to somehow handle and process safely.

What of the new generation of reactors and future favourable economics? One of the technologies much vaunted by the industry involves a so-called pebble-bed reactor. Such reactors use many thousands of spherical fuel assemblies, each of which consists of an inner graphite core embedded with thousands of smaller fuel particles of enriched uranium encapsulated in multiple layers of hardened carbon. They are under development by South African utility company Eskom, but have yet to prove they can operate commercially without the problems earlier reactors have had. Indeed, a trial pebble-bed reactor in Germany leaked radiation when a pebble became lodged in the pipe feeding fuel to the reactor just before the Chernobyl accident in 1986. It was shut down permanently.[294] Watch carefully to see how often nuclear-industry advocates own up to this. As for the nuclear advocates' arguments about nuclear's ability to compete in liberalized markets, be aware that the low costs the industry likes to quote come from plants where massive debts have been written off, and the capital costs of the new plants will remain high: $2,000 per kilowatt of electric capacity, according to the International Energy Agency, compared to around $1,000 for coal, even before government subsidies and waste-disposal costs are factored in. Regarding the latter, consider the recent experience of British Energy, the main UK nuclear generator, which is supposed to have set aside capital for waste disposal. In 2002, the UK government had to pay £2 billion of public money, with more to come in the future, to stop the company being bankrupted by waste-disposal bills.

Finally, another thing always to bear in mind with nuclear power is the scope for human error. This applies to any energy technology, but in the case of nuclear the consequences of a failure are awful indeed. Think of the engineers whose games with the controls at the number four reactor at Chernobyl went so wrong on 26 April 1986. Can we expect that Russian engineers would be the only people prone to such random lunacy in a world of proliferating nuclear power? I think of firefighter Vasily Ignatenko, who was thrown into the carnage at the plant as the Soviet authorities fought to contain the damage. His wife Lyudmilla describes his subsequent slow death in hospital. "He started to change," she recounts. "Every day I met a brand new person. The burns started to come to the surface. In his mouth, on his tongue, his cheeks – at first there were little lesions, and then they grew. It came off in layers – as white film ... the colour of his face ... his body ... blue, red, grey-brown. And its all mine!" She watched grief stricken as her husband died in this way over the course of fourteen long days.[295]

Who needs to risk this kind of thing happening again, even to one more human being, whether from another low-risk, high-consequence case of human error like Chernobyl, or from the detonation of a nuclear device made possible by the existence of nuclear power stations? Why should it be necessary when nuclear power itself is unnecessary?

A functional multilateral world with effective international law?

Suppose we have some warning of the approaching oil topping point. Matthew Simmons has pointed to transparency over oil reserves as a key requirement. "I think we should basically look at this like we looked at nuclear warfare and say that that would be so awful if it happened – let's do something, put in a warning system," he urges.[296] Suppose governments listen to such entreaties. Even then it is difficult to imagine a smooth transition from oil addiction to alternatives of any kind.

Colin Campbell has proposed a greenhouse-gas depletion protocol, something that the town of Rimini in Italy is planning to adopt. A Rimini Protocol might, for instance, require businesses to cut imports to match the world depletion rate of around 2 percent. A 2.5 percent-a-year reduction in imports is attainable, Campbell argues. The result would be impressive: modest prices that even poor countries could afford, minimal energy needs met, and profiteering avoided. Best of all, consumers everywhere could be educated and prepared to face the reality of a changing world.

Is there a chance of such selfless collective thinking breaking out in the field of international relations? Given the experience of the Climate Convention, with its decade-plus of status-quo defence by vested interests, and the state of American politics, where the Bush Administration's energy plan amounts to a dependency pact, I am forced to conclude that the prospect is at best a very slim one.

The most probable outcome

The most likely outcome is that the world will drift on in overall collective denial. This book will be published, to join the other efforts under way to wake people up. It will be praised by some, vilified by others, but ignored by most people. At some stage, as the topping point approaches, there will be a gathering upsurge of speculation that maybe the early-topping-point whistleblowers are on to something. Then the tsunami will hit.

The tsunami might be one giant wave of unstoppable panic and tumbling markets, like the Great Crash of 1929, or more than one wave, none giant but cumulatively as bad or worse than a single giant. In the latter scenario, the oil price gets to a point that triggers recession. In that recession, as economic activity shrinks, the low-hanging fruit of energy savings get picked as soon as possible. Demand goes down as a result, so the oil price goes down, so the economy improves. But demand for oil then goes up with the improving economies, and supply once again is inadequate – or perceived to be so – so the price goes back up and the economy dives again. The tsunami might therefore be a cyclical phenomenon of closely spaced oil-price peaks and recessions until the point that there is no more low-hanging alternative fruit, and the long-run shortfall between energy demand and supply becomes clear.

Major world events completely unrelated, or only indirectly related, to the oil topping point will inevitably muddy this picture. There are more reasons than the price of oil for global markets to crash, as George Soros often points out. It is easy to think of possibilities that might intersect with the playing out of the oil-depletion drama in ways that could accelerate or delay, amplify or suppress, a fusing of the psychological peak panic point with the physical oil-production topping point. An acceleration/amplification might be the fall of the House of Saud to a fundamentalist regime. A delay/suppression might be a major setback for the Chinese economy. I am not going to speculate here about the detail of how the crisis will play out, or what the years immediately after the tsunami – whatever form it takes – will look and feel like sector by sector across society and economy. Suffice to say that it doesn't look very pretty, however you look at it.

The main point is this. In the wake of the realization of the oil-production topping point, the onus in the energy sector will be on immediate damage limitation. Energy services and renewable energy companies will be besieged. Politicians will want to know how quickly we can accelerate if they finally give us as much support as they have given nuclear and the military pursuit of oil-supply protection all these years. Corporations will want to cut deals with us exclusively to keep their lights on and not their competitors'. Consumers will suddenly be desperate to be taken off the national grid and given heat and electricity at almost any price. The eyes of the world will focus on Woking, as it were. Come whatever in other societal and economic sectors, people working in renewable energy, energy storage and energy efficiency will be in the front row of those who can help once widespread acceptance of the oil topping point and its implications has descended on the world.

3. RENEWABLE ENERGY AND FUEL USE, ALONGSIDE ENERGY EFFICIENCY, WILL INCREASINGLY SUBSTITUTE FOR OIL AND GAS, GROWING EXPLOSIVELY WHATEVER HAPPENS

The first major financial institution to set up an investment fund for renewable energy suffered something of a disaster. Merrill Lynch's £200 million New Energy Fund floated shortly before the dot.com crash of 2000 and was one of the many victims trailed in its wake. This was a painful experience for many people, including me: I put my meagre life savings in that fund. Big money was put off for years. But in 2004, with the dot.com bear market over, oil prices rising, and the prospect of the Kyoto Protocol coming into force, a number of major financial institutions began setting up investment funds for renewable energy, and that trend has accelerated into 2005. Just as in the run-up to the dot.com boom, the business magazines finally began to take notice. *Business Week* asked on its front cover: "Global warming: why is business taking it so seriously?"[297] *Fortune* went so far as to draw up its own plan for a renewables revolution.[298] Veteran investors began to talk of the dangers of a boom in renewables investment, and even a dot.com-style feeding frenzy.

I have seen much of this change in interest among the serious money people first-hand, since serving as a founding board director of the world's first private-equity renewable-energy fund.[299] The solar PV industry serves as a good example. The journal *Photon* records an index of fourteen quoted companies whose businesses are more than 50 percent solar PV. In 2004, this index rose 182 percent. In contrast, and notwithstanding a persistently high oil price, the index for the twelve largest oil stocks rose only 18 percent.[300] Of course, the fourteen solar companies have a collective market capitalization of not much more than $1 billion, compared to goodness knows how many hundreds of billions for the oil giants, but investors like growth. That is where they go first for investment options.

The Photon index excludes the solar companies with giant parents doing mostly other things for the moment, such as Sharp and Sanyo in electronics and BP and Shell in oil. It also excludes all the private companies. When all these are added in, the embryonic PV industry doesn't look so embryonic after all. In July 2004, Crédit Lyonnais Securities Asia conducted the most exhaustive sector study yet of the PV industry. They found a market of more than $7 billion growing fast at 30 percent per year, and an overall profit pool of $800 million. They expect the industry's growth to reach $30 billion in revenue and more than $3 billion in profit by 2010. As they worked their way through more than two hundred solar companies, the CLSA analysts started out sceptical and ended up enthusiastic. In their writing, they can barely contain their surprise. "Lost in the noise has been a deeper, fundamental story," they conclude. "Solar power is hot." [301]

One reason for this is that solar power can compete with retail *prices*, not generator *costs*. This is an important point to bear in mind when it comes to personal or collective action in the coming energy crisis. Solar PV is a unique technology, generating right where the power is needed – primarily in buildings – and so bears direct comparison with the retail price of power, including everything the utilities load into those prices: their generating costs in conventional polluting power stations, transmission and distribution costs in the grid, taxes, exorbitant profits and anything else they can get away with. Yet time and again energy analysts compare the costs of a PV system with only the *generating* costs of coal, nuclear, gas, oil or wind. No wonder people say solar is "too expensive" (three to ten times more so, on this unfair playing field). The CLSA analysts have one word for this: "irrelevant".

Another reason for the soaring investor interest in solar is that PV costs are coming down at 5 percent per year on average over the last few years, while the price of oil and gas continues to search for the roof. Solar manufacturers have significant economies of scale to look forward to as they scale up. Also, they apply knowledge gleaned at low volumes to achieve efficiencies at higher volumes. For this reason, costs decrease by around 20 percent for each doubling of capacity – a long-term trend that is expected to continue in the years ahead. In fact, there is potential for costs to come down even faster. Meanwhile, what is oil doing but generally up? And where oil prices go, gas tends to follow.

A further reason is that solar power is already more economical than the polluting alternatives in many markets. In Japan and Germany, for instance, the two largest markets for PV, government market-building incentives have made the price of solar power competitive with the residential grid power price by, respectively, applying direct subsidies and using a premium buyback rate. The CLSA analysts conclude: "Our initial reaction to solar's dependence on incentives was to discount the potential of solar power. Our view has changed as we became more convinced that incentives result from perceived global climate change risks and energy security/price concerns that are unlikely to disappear anytime soon." Moreover, the Japanese subsidies are now being phased out, and yet there is no sign of the market slowing.

This sort of interest in the money world for renewables is real and growing. It is causing the renewables industries to explode up hockey-stick growth curves. They can become very big, very quick. And the longer it takes for the full weight of the energy crisis to hit, the better they will be able to limit the damage.

But this burst of optimism has to be severely tempered. The oil and gas industry is vastly bigger. So too is the coal industry.

4. AMID THE RUINS OF THE OLD ENERGY MODUS OPERANDI MANY WILL TRY TO TURN TO COAL, AND SO THE EXTENT TO WHICH RENEWABLE ENERGY GROWS EXPLOSIVELY INSTEAD OF COAL EXPANSION, RATHER THAN ALONGSIDE IT, WILL DETERMINE WHETHER ECONOMIES AND ECOSYSTEMS CAN SURVIVE THE GLOBAL-WARMING THREAT

The coal industry is strangely hardline. It utilizes a technology that is so clearly mortgaging the future – at best, torpedoing it at worst – and yet it continues to grow largely unapologetically around the world. Beside the future death toll from unmitigated global warming and dire air quality, there is also the actual to-date death toll from getting the stuff out of the ground. Type "coal-mining disasters" into Google and see what I mean. The people who run coal really can't care that much about their workers, can they? In the face of all this, even if they genuinely reject scientists' arguments about global warming, you would think there would be a little humility, a little reluctance about the product, a little willingness to search hard for alternatives. No, not that I have seen. In all my years as an environmental campaigner, I have observed a clear distinction between the oil industry and the coal industry. It is difficult to generalize about cultures across entire industries, and I would hesitate to do so if the impression had not been instilled in me by considerable on-the-ground experience. The oilmen and -women were capable of politeness, reasoned debate, and even changing their position. The coal men and women weren't; not that I ever saw, once.

Many examples spring to mind. In May 1993 I addressed a joint meeting of the US National Coal Association and the European coal industry in Barcelona. I shared a platform with an American who believed environmentalists were defeated communists who had turned green, and told the listening coal executives so. The flavour of his language suggested that we were opportunistic ecoterrorists working to bring down capitalism. He exhorted the executives to mobilize against us and get as much coal to market for burning as quickly as possible. I made an effort to persuade him he was wrong. I told him about the concerned professionals who were jumping ship to Greenpeace in growing numbers at the time, doing their best to bolster the credibility of the idealistic kids and gentle veterans. I told him how proud I was never to have witnessed a transgression of Greenpeace's code of non-violence. I told him how some of my colleagues were actually quite conservative in their political views, and how a public party-political statement was a serious offence in the organization. I remember his face as I spoke: the blankness in his eyes, and the twitch of a facial muscle as he framed his dismissive response. I am talking to an idealogue, I thought, a man who needs his enemies to be as he imagines them. That man's name was Harlan Watson. He is now chief negotiator for the United States of America at the international climate negotiations.

What does Watson have to say for himself these days on the subject? "There is a growing realization that existing energy technologies, even with substantial improvements, cannot meet the growing global demand for energy while delivering the emissions reductions necessary to stabilize atmospheric greenhouse-gas emissions concentrations." Hey, some improvements then, after more than a decade. But wait. What are the implications? "We need to develop and deploy globally 'transformational' technologies — that is, revolutionary changes in the technology of energy production, distribution, storage, conversion, and use. Some examples include carbon sequestration, hydrogen, and advanced nuclear technologies."[302] No mention of renewables or efficiency. The key word here is "sequestration", and it means that even when he talks about hydrogen he means with coal gasification as the primary source.

The sad fact is this. Watson speaks for a huge constituency in the coal industry, and their supporters in governments, that is endeavouring to find ways of capturing carbon dioxide somehow, either at the power plant or in the air, and storing it somewhere so that it doesn't trap heat. These people will gladly throw billions of dollars at such sequestration techniques before they think about exploring the family of renewable and efficient technologies.

A second category of coal defenders don't care anyway about global warming, and when the oil production tipping point hits will surely – based on their record to date – exhort the world to burn away merrily. America is likely to lead the charge. It has the world's biggest reserves of untapped coal, with two hundred and fifty years' worth of supply on current demand levels. As things stand, at least ninety-four coal-fired electric power plants are in planning across thirty-six American states. These proposals have piled up in the three years since the Bush energy plan rolled out. They would add 62 gigawatts, or another 20 percent, to the US's current coal-generating capacity, which meets half the nation's electricity demand as of today. That said, not all of them will be built. It can take seven to ten years for a coal power plant to go from planning to construction. The track record shows that legal action and public protests, like those facing the nuclear industry, often halt them.[303]

Many of the people predisposed in America to burn coal come what may will inevitably be fundamentalist Christians who believe in the "rapture". Their view is that the bible predicts an environmental holocaust, after which the faithful will ascend to heaven: ergo, burn away and roll on the day![304] (How many fundamentalist rapturists work for the Exxons and Peabody Coals? Therein lies a worthwhile research project for an investigative journalist.) Others who care not a jot about greenhouse gases, and will happily burn hundreds of billions of tonnes of coal, include libertarians and neoconservatives. Michael Crichton's novel *State of Fear* gives an insight into this state of mind, in the sense that he reveals himself in it to be one of them.[305] That said, many non-fundamentalist non-idealogues will also advocate a "burn and be damned" policy. In China and India, countries with both exploding energy demand and lots of coal, there may be a common perception that there is just no choice. As things stand, the Chinese are building a gigawatt of new coal capacity each week.

On top of all this will come the clarion call, no doubt in many countries, for gasoline to be made by coal gasification. People will be looking anywhere they can for gasoline substitutes once the oil topping point is upon us. Squeezing oil out of coal is one method for doing that, costly and environmentally disastrous though it may be.

The prospect of widely attainable sequestration technologies will deepen the temptations to go all-out for coal. The discussion in Chapter 7 shows what will happen, on the massive balance of probabilities, if we do. If there is even a slight mismatch between the amount of carbon burnt in coal, and the amount of carbon in carbon dioxide emitted instead of sequestered – as there surely would be given the scale of the current demand and the embryonic state and riskiness of the sequestration technologies – we would soon burst through the danger threshold of a 2°C increase in the global average temperature and soar beyond. Then, to put it decorously, we would find out if the climate scientists are right.

Sequestration of carbon dioxide from coal

Let us take a quick tour of current research and development in sequestration to give a feel for the scale of the challenge facing advocates of renewable and efficient-energy technologies when it comes to the matter of coal. Proponents envisage three types of sequestration: geological (the injection of carbon dioxide into rock strata below ground), oceanic (the pumping of carbon dioxide into the oceans) and biological (the stimulated soaking up of carbon dioxide by plants onland or in the sea).

When it comes to geological sequestration, the United States is leading the way. As US Energy Secretary Spencer Abraham put it in 2003, "... carbon sequestration has rapidly grown in importance to become one of the Administration's highest priorities. Our activities and our plans bear out the determination with which we are pursuing the promise." Accordingly, the US Department of Energy's sequestration programme has several active research projects under way. In Canada, New Mexico, Virginia and Texas, carbon dioxide is already being injected, or will be injected if the plans go live, into one or more of the three main underground storage options: depleted oilfields, unmineable coal seams and deep saline reservoirs. By 2009, the DoE's aim is to initiate at least one large-scale demonstration of carbon-dioxide storage (>1 million tonnes per year) in a geological formation. Estimates of sequestration costs based on the technologies available today are in the range of $100 to $300 per tonne of carbon emissions avoided. The goal of the sequestration programme is to reduce the cost of carbon sequestration to $10 or less per net tonne of carbon emissions avoided by 2015. By 2050, the Department professes that this technique will be saving more carbon-dioxide emissions than renewables and efficiency combined.[306]

The European Union is collaborating enthusiastically with this research, and has significant numbers of programmes of its own in the works. The UK government has produced a report saying that large-scale sequestration may be needed to help it reach a target of 20 percent greenhouse-gas emissions' reductions by 2020. However, in Europe at least some scrutiny of the potential downsides of sequestration is likely. The same report concluded "… it is currently impossible to analyse with any confidence the likelihood of accidental releases from carbon dioxide sequestration reservoirs".[307] It further raises questions about the legal liability of companies if gases do escape, and the insurability of operations on a large scale. But hey, what would a little gamble matter to someone desperate for some unfathomable reason to save his or her filthy and lethal industry when there are perfectly acceptable clean alternatives of many types?

Proponents of oceanic sequestration argue that carbon dioxide from power plants could be piped across land, and the continental shelves, and then pumped into the deep oceans. The argument is that the gas will dissolve in sea water at depth, and not be returned to the surface for hundreds of years.[308] Such an approach would contravene the international treaty controlling waste disposal at sea, the London Dumping Convention. Moreover, major concerns about acidification of the oceans ought to render this idea a non-starter. It appears from the most detailed measurements yet made, published in 2004, that the oceans have absorbed almost half of all carbon dioxide from fossil-fuel burning and cement manufacturing to date. "The oceans are producing this tremendous service to humankind by reducing the amount of carbon dioxide in the atmosphere," says Chris Sabine, an oceanographer at the US National Oceanic and Atmospheric Administration in Seattle who has been a leader in this research. "But it's changing the chemistry of the oceans." The research results are based on ten years of criss-crossing of the globe by ships making nearly 10,000 stops for measurements along the way. The core problem is that carbon dioxide dissolves in sea water to form carbonic acid, which in turn can dissolve the shells and skeletons of marine life.[309] Great. We can carry on burning coal, but we have to do without food for the world's fish, which give humans most of their protein.

Beyond geological and oceanic sequestration, others have proposed an all-out effort to boost biological sequestration. We have known that carbon dioxide can be taken out of the atmosphere and into fast-growing plants for as long as photosynthesis has been understood. A hectare of immature forest can absorb more than 100 tonnes of carbon each year.[310] But recent detailed studies have shown that the effect is not as significant as biologists once thought. Certainly it is nowhere near big enough to offset emissions from fossil-fuel burning.[311] Undeterred, some scientists have proposed boosting the amount of carbon dioxide taken out of the atmosphere by increasing the numbers of phytoplankton, single-cell plants that float by the myriad in surface waters of the world's oceans. This could be done, they say, by doping the oceans with iron, so boosting nutrient for the phytoplankton, causing them to bloom.[312] This is yet another example of a proposed reckless gamble in a poorly understood and visibly stressed environment, when perfectly safe alternatives are available in volume in renewable and efficient energy technologies.

How scientists can easily accelerate a rush to coal

Such willingness by scientists to gamble with the natural environment extends even into space. High-tech, sci-fi-like schemes for cutting down the amount of solar radiation hitting the surface of the planet include the idea of billions of metal-coated reflective balloons in the stratosphere – a sort of "optical chaff" aiming to bounce incoming solar radiation back into space – and giant orbiting reflective mirrors.[313] As they enthusiastically haul their plans off the shelves in the wake of the oil topping point, these techno dreamers will also indirectly help the case for accelerated coal use.

Why do scientists fly off on such risky or even crazy tangents when the well of real-life renewable-energy, energy-storage and energy-efficiency alternatives is so deep and productive? I have long wondered about this question. I taught and researched for eleven years at one of the world's top science and technology universities and I know a little about the scientific community. I would offer two observations, neither or both of which can presume to be a complete answer. First, the culture of science is rarely holistic. Practitioners are encouraged to achieve excellence in one discipline, and often one small sub-sector of one discipline. Being an all-rounder is not seen as a route to greatness. Being an all-rounder is in fact actively frowned upon. For this reason, scientists tend not to be schooled in the kind of all-embracing view that a total-ecosystem approach to risk analysis demands. How many nuclear physicists fifty years ago – the people who told us that they could give us safe electricity too cheap to meter – knew anything much at the time about the geology involved in disposing of their waste? How many genetic engineers today – some of the fiercest advocates of massive releases of genetically manipulated organisms into the environment – know anything much about ecology? You see my point. So if you know about space and satellite technology, and you hear there is a danger of global warming, you might well be tempted to advocate orbiting mirrors rather than CHP, PV and smaller electricity grids on Earth. And because you are a famous exponent of your craft, maybe even a Nobel Laureate, you are actually in danger of being taken seriously. Second, institutional science tends to encourage individualism. The reward system is all about lionizing the stars. This encourages a tendency to personal arrogance in the leaders of the profession that, in my experience, tends not be found so often in their equivalents in the business world.

With that thought I end my survey of the forces which, directly or indirectly, will favour a massive retreat into coal when the panic descends in the wake of the oil topping point. If that point is later in the decade rather than imminent, the sense of opportunity and belief in the feasibility of renewables, storage and efficiency as alternatives may be much greater than it is today. But then again – given the research programmes under way aiming to bolster the acceptability of coal in a warming world, the depth of denial in the coal and coal-related theocracies, and the scope for iconoclastic scientists to fan the flames of confusion – belief in the feasibility of burning a thousand billion tonnes of coal or more and getting away with it might be much greater too.

The bottom line is this. Two ideas will confront each other head-on in the wake of the oil topping point. We can call them solarization and coalification. This, I contend, will be the battleground that will decide the fate of the planet.

5. THERE IS MUCH THAT PEOPLE CAN DO TO INFLUENCE THE OUTCOME OF THE STRUGGLE BETWEEN SOLARIZATION AND COALIFICATION, TO AMELIORATE THE WORST EXCESSES OF THE GLOBAL ENERGY CRISIS, AND TO CREATE A BETTER SOCIETY IN THE PROCESS

Leadership by government?

Can we wait for governments to sort out the mess? Looking at the history of the Kyoto Protocol offers only a miniscule amount of hope. So much time spent negotiating, and so little to show for it. While fifteen years of talks have produced a mechanism that may in time freeze emissions overall in most developed countries, absent America of course, how have the fossil-fuel industries fared subsidy-wise? Annual government subsidies to oil, coal and gas total $235 billion just using conservative assumptions, according to the New Economics Foundation.[314] That would make $3.5 trillion in handouts, in round figures, since the climate negotiations began. In the cases of almost all the governments involved in this largesse, their scientists are telling them that fossil-fuel addiction will present their successors with huge bills, perhaps even unpayably huge bills. Why do they not cancel even a few of these subsidies and pump the money into alternatives? If we do have time to make the shift from hydrocarbons to alternatives without too much economic disruption, the list of upsides for governments would be impressive. For many of them, the technologies of the solarization era hold the potential for an escape from foreign oil and gas dependence, and even war. More than that, the air would be cleaner, so that rates of lung cancer, asthma and other respiratory diseases would drop. Governments could then spend less on healthcare and more on, say, education. Local energy supply would build local economies and therefore jobs, leading to the revival of communities of many kinds and cuts in unemployment. Renewables on coasts would allow desalination on a massive scale, averting many growing and potentially horrific problems of water supply. Addressing or even solving those problems would in turn improve or solve many agricultural problems, increasing governments' chances of feeding their populations. Local electricity in the developing world would allow the refrigeration of medicines, and the pumping of clean water, with major impacts on child mortality. With all these improvements, social unrest would surely ameliorate. The benefits would roll on. Indoor pollution accounts for 2.2 million deaths and between $150 and $750 billion in lost production resulting from sickness and death. All of non-electrified sub-Saharan Africa could be provided with energy from small-scale solar for less than 70 percent of what the wealthy OECD countries spend on subsidies for fossil fuels every year.[315]

So often in these discussions the focus is on the developed world. But at the community level in the developing world the opportunities are vast. Many studies have shown that fully 400 million households in developing countries could afford solar electricity if the channels of microcredit and distribution could just be put in place on a widespread basis. Those households are already paying more as things stand for sources of power like candles, kerosene, generators and dry cell batteries.[316] It has taken multilateral financial institutions like the World Bank – themselves controlled by governments – far too long to accept that investment in channels of credit are vital and low risk, vulnerable as they are to only tiny default rates.[317] But are governments doing this simple, blindingly obvious, thing? Some wanted to at the Johannesburg Earth Summit in 2002, but the US didn't, and the others were easily sidetracked by the horse trading involved at the end of the summit. No, governments cannot be relied on to take the lead in the modern world. Many of them are led themselves, by industry. They will need to be pushed.

Leadership by the law?

When faced with unassailable evidence that something major has gone badly wrong, society has a habit of looking back in anger. The tobacco industry is currently the embodiment of this phenomenon. Until the emergence of the first evidence that cigarette smoking causes cancer, the industry was able to project an image of unsullied glamour. Now, it faces the prospect of lawsuits without end. It stands accused of wilful and lethal harm to consumers and non-consumers alike, and criminal attempts to manipulate and suppress the evidence that it might have a problem with its core product.

In recent years, lawyers have turned their attention to climate change with thoughts about liability somewhat akin to the case of tobacco. The first case of this kind was launched in 2004 when New York's campaigning Attorney General, Elliot Spitzer, and seven other states sought a court edict imposing cuts in carbon-dioxide emissions on five major American energy corporations. The companies facing this new legal frontier in the fight against global warming are American Electric Power Company, Southern Company, Tennessee Valley Authority, Xcel Energy and Cinergy Corporation. Together, they own or operate one hundred and seventy-four fossil-fuel-burning power plants in twenty states that emit some 650 million tonnes of carbon dioxide yearly, about 10 percent of the US total.[318]

An international network of lawyers, run by my friend, ex-Shell barrister Peter Roderick, is working to bring cases against oil and other fossil-fuel companies on a broader front, including redress for the years in which the Global Climate Coalition and other Carbon-Club organizations have obfuscated and worse on behalf of fossil-fuel companies at the climate negotiations. Two factors make eventual success for the legal campaigners increasingly feasible. First, scientists are becoming more willing to assign probabilistic causes to extreme climatic events. For example, the European heatwave of 2003, which caused 14,000 excess deaths in France alone, was so far outside the range of precedents that scientists feel the link to greenhouse gases can be quantified at the 90 percent level.[319] Moreover, since 2001 three ground-breaking peer-reviewed papers have attributed human influence on regional temperatures. So-called "detection-and-attribution science" is moving forward, and this is of profound importance to the lawyers waiting their chance.[320] Second, and worryingly for the companies concerned, a substantial fraction of the increase in atmospheric greenhouse-gas concentrations in the post-war years can be traced to products produced, sold or used by only a few dozen companies.[321] It is this focus on complicity that allowed the Spitzer lawsuit to be brought to court, and it may yet enable many similar cases to be heard.

Cause for concern should extend well beyond the prominent few dozen, into the big financial institutions. It is surely only a matter of time before the first legal cases are brought against investors, for being asleep at the wheel when it comes to fiduciary responsibility: their legal responsibility to shareholders to invest wisely on their behalf. A Lloyd's insurance broker has suggested the following scenario to me. A shareholder collects every scientific report on climate change over the last decade and sends them in a package to every director of an insurance company with a letter asking what the company is doing to safeguard his or her investment in the company against the threat of climate change. The answers, as things stand, would have to be contortedly evasive and would potentially arm the shareholder with material for a very interesting lawsuit.

And what of depletion on this basis? Will society be any less inclined to look back in anger if an early topping point does in fact materialize? Will there not be some very close scrutiny of all the optimistic assertions that there was plenty of oil left, if in fact there proves to be nowhere near enough?

As BP CEO Lord Browne writes comfortingly in the foreword of BP's most recent Statistical Review of Other People's Data: "… at current levels of consumption there are sufficient reserves to meet oil demand for some forty years." [322] This has to be, at minimum, a risky thing to say. "Sufficient reserves" includes the OPEC add-ons Browne must know to be a figment of OPEC imaginations. Colin Campbell says of Lord Browne's statement: "… this goes beyond being economical with the truth, and suggests a deliberate policy of denial and obfuscation, which deserves to be exposed. The most fatuous and misleading approach is to take the reserve number and divide it by current production to say that the reserves support current production for forty years, ignoring the natural decline observed in all fields and countries. It is in this regard that BP deserves serious criticism. If its objective is to evade the issue of depletion to impress the stock market with the pretence that finding oil is just a matter of economic incentive and technology, it risks being accused of culpable fraud." [323]

One person's view, perhaps. But if I were Lord Browne I would worry about this. The more so because an acid test will be how much investment the industry puts into exploration in the years ahead. As we have seen, according to the likes of Goldman Sachs and the International Energy Agency it must be lots, much more than the industry has ever invested before: almost $250 billion per year over the next ten years if future demand is to be met.[324] If the oil companies really believe we are running into oil, not out of it, they will go ahead with investment of such enormous sums and do so with gusto, especially with the levels of profit they are raking in at today's record oil prices. But if, in their hearts, they don't believe enough oil is there …?

Here is my prediction. Law firms are going to be hiring geologists and accountants in the years to come.

Leadership by corporations?

What about companies? Can the progressives in the corporate world lead the way to sanity and survival? At the World Economic Forum in Davos in February 2000, several hundred CEOs of the world's biggest companies were asked by the organizers to vote on the greatest challenge facing the world at the beginning of the new century. Global warming came out top. The organizers were so surprised they tried the exercise a second time, fearing a mechanical failure. That time the score for global warming was even higher. What does this tell us? Are the CEOs of the world's biggest corporations about to leap up on the barricades with environmental campaigners? No. But at minimum it surely means that the business world is well primed, and that a little concerted consumer pressure can go a long way. I believe it means more, though. I sense that we are living through turnaround years, and that corporations are beginning to suspect that their own prosperity and survival lies in doing something meaningful about the pressing environmental and social issues of the day. Recently I watched a senior executive from Toyota interviewed on TV. The company's results had just come out, and unlike most auto manufacturers' they were excellent. Toyota's market capitalization was around four times that of General Motors and Ford. The interviewer asked whether the company's ambition was to replace

General Motors as the number one manufacturer, in terms of turnover. We want to be number one, replied the executive, but not in size. We want to be number one in the respect of the world.

I think also of HSBC, one of the world's biggest financial institutions, which while I was writing this book announced that its entire global operations would be made carbon neutral. It is not before time that a bank or insurer made such a meaningful move. As long ago as 1996, the first two books on the financial sector and the environment, drafted by, among others, senior figures from the financial sector itself, called for urgent action. The World Business Council for Sustainable Development's book, *Financing Change*, took the bigger view, asking the question "Are the world's financial markets and those who work in and around them a force for sustainable human progress, or are they an impediment against it?" The answer, paraphrased, was: an impediment, but with the means and the business imperative to change. (As one of the drafters, a senior member of one of the banks in the WBCSD, told me, "The book should have been called *Changing Finance*.") The book urged company leaders to be proactive. It exhorted investors to find ways of pricing sustainability and to provide information to shareholders, such as pensioners, so that they better influence those who make decisions on their behalf. It suggested that bankers find ways of providing financial products and services that reduce environmental risk, and that insurers invest as though climate change was real.[325] The second book, which I edited, went to town on that last point.[326] The main message of the two books remains an unrealized dream: that the financial institutions invest in survival, not suicide. But that dream is closer to potential fruition today than it has ever been, as HSBC's carbon-neutral decision shows. Reaching the dream isn't guaranteed, but it surely is a tantalizing possibility, as things stand.

One of the crucial battlegrounds will be shareholder activism against the oil companies. In February 2004, several of those companies faced a record number of shareholder resolutions at their AGMs. Significant numbers of major institutional investors – among them four state and city pension funds, a foundation, socially responsible investment firms and a number of religious pension funds, with a collective representation of over $250 billion in assets – supported these resolutions, even though they originated from shareholder activist groups. Most of the resolutions sought reports on how the companies in question were responding to and preparing for rising regulatory and competitive pressures to reduce greenhouse-gas emissions. Resolutions for ExxonMobil, for example, requested reports on the company's investments in renewable energy (answer: desultory of course) and full disclosure of the science supporting the company's policies on climate change (honest answer: lies and obfuscation). As an oil-industry analyst put it: "The disparity of pre-paredness among the companies is disturbing. All oil companies essentially operate in the same global markets and are susceptible to the same emerging regulatory structures around the world, yet many of these companies seem relatively uninformed about the issue and how it could affect prices. It seems that US intransigence on global warming has translated into insularity that puts US companies at serious risk." [327]

Where could all this be leading? Just how fertile might the ground be for consumer pressure on the big energy companies to lever seismic change? Here is a thought for marketing directors to mull over as the oil topping point approaches. One of the biggest advertising agencies in the world, Young and Rubicam, has declared that belief in consumer brands has replaced religion in people's lives. "People are turning to them for meaning," it says. "The brands that are succeeding are those with strong beliefs and original ideas. They are also the ones that have the passion and energy to change the world, and to convert people to their way of thinking through outstanding communication."[328]

There, surely, is something to shoot for. Ideas can and have taken off on the scale we will need if we are to come out the other side of the oil topping point with solarization, not coalification, in our gunsights.

People and tipping points

When such a take-off happens, the process will be like an epidemic. So argues Malcolm Gladwell, a former business writer on the *Washington Post*, in his inspirational book *The Tipping Point*.[329] The types of ideas Gladwell talks about range from individual consumer products, individual companies, campaigns of all sorts, revolutions business or otherwise. The principle is general. The tipping point is where the idea stops growing slowly and reaches a critical mass beyond which it accelerates explosively. There are three particularly important rules governing whether any idea will "tip", Gladwell argues, all drawn from medical understanding of what makes an epidemic spread. He calls them the law of the few, the stickiness factor and the power of context.

The law of the few holds that a tiny percentage of people make an idea tip. One classic example, trivial as it may seem, involves Hush Puppies. For many years after their initial success these shoes were consigned to "old fogey" status, and found only in unfashionable shops. Then a few kids started wearing them as fashion statements in Manhattan and within a few years every American boy owned a pair. Another example is Sharp's commercialization of the fax machine in America. They started in 1984, with a big barrier: you need to know other people have a fax machine before getting one yourself. Sales grew steadily until enough people had a machine for the explosion to take place, which it duly did in 1987.

The "few" can be thought of as connectors, mavens, and salesmen. Connectors have a special gift for bringing people together. They have huge networks of "weak connections" ideal for the spreading of ideas. The e-mail makes this kind of person particularly important these days, a fact that viral marketing exploits. Mavens are people who love accumulating information, the type of people that tend to be looked up to and listened to. Any campaign involving technology must get to the mavens. They provide the message, and the connectors spread it. Salesmen are the types who persuade the unconvinced. Whether they are directly selling a product or idea, or indirectly supporting one because they happen to believe in it, they are people with subtle skills in persuasion. If the connectors can reach them, they in turn will swing many more people.

Like the more effective viruses and other bugs, the message has to be "sticky" in order to spark action. Many ideas simply drift by people because they don't stick. One of the much-studied success stories mentioned in the book is *Sesame Street*. The muppets and monsters were used to make educative messages stick with children, and they succeeded hugely. Things do not seem to be much different with adults and advertising.

Perhaps the most interesting idea in Gladwell's book involves the power of context. People tend to analyse events involving others in terms of fundamental character much more than by situation and context. But the book argues that we are wrong to do this. A key example is crime in New York City. Working back from the criminologists' "epidemic" theory of crime, which holds that if disorder is evident – run-down neighbourhoods with broken windows unrepaired, say – crime will spread, the City of New York began tackling litter, broken windows, and minor misdemeanours such as dodging fares on the subway. Within a few years the crime rate – up to and including the sky-high murder rate – had plummeted. Another key component in the power of context is the role of the small group. The size of a group hugely affects its receptivity to ideas and Gladwell illustrates this by use of a fictional book about a group of women, *Divine Secrets of the Ya-Ya Sisterhood*. The book itself tipped hugely by word of mouth in small regional meetings of women's community groups. Once the epidemic was under way, women all over America started forming Ya-Ya Sisterhoods just like the one in the novel.

There are many thought-provoking ideas and case histories in Gladwell's book, all of them relevant to the playing-out of the great global energy crisis and its aftermath. The sum of the core ideas is ultimately very hopeful. By tinkering with the presentation of our messages, we can make them sticky, and so reach many people with the pulling power to shape social epidemics. Then we can expect corporations, governments and institutions to act. As Gladwell says in the last words of his book, "Look at the world around you. It may seem like an immovable, implacable place. It is not. With the slightest push – in just the right place – it can be tipped."

Incitement to action: enlightened self-interest

With Gladwell's ideas in mind, how exactly might people contribute to a tipping point for solarization, so that the idea of viable alternatives to oil and gas in renewable energy and energy efficiency grows far faster than the idea that we should mire ourselves ever deeper in coal?

As individuals, we need to remember Ida Tarbell, and resurrect her spirit. As I recounted in Chapter 6, Ida was the journalist whose investigations and writings brought down the forerunner of Exxon, Standard Oil, in 1911, causing that thuggishly un-American monopoly with its anti-social practices to be divested, or dismembered as the oilmen thought of it. Individual people can and have had impacts on these vast corporations. It nearly happened again in 1974, when those forty-five senators voted for divestiture of the oil majors in the wake of the first oil crisis. The eight hundred potential bills concerning energy that circulated in Congress at the time, most of them favouring early versions of the smart solarization technologies that are so attractive to venture capitalists today, amounted to little, as we saw, but the important thing is that the oil companies were at that point willing in principle to diversify out of oil, and they will be again. Come the third oil crisis, we must not let them off the hook so easily. BP and Shell may be far more civilized today than they ever were in the twentieth century, but Exxon is still much the same ugly entity it has long been, and all of them need to be diverted substantially from the course they are on. We live in an age in which it is easy for mavens to access information and organize it into messages, and for connectors to organize networks into which messages can be spread by salesmen. This is the reason that corporations live in such fear of consumer power. One vast upsurge of consumer anger against a corporation can ruin it. This, not an intrinsic desire to save the planet, is probably what is motivating most of the progressive companies today. It would be a mistake for an individual to do nothing because he or she thought that anything doable would have no impact. Small, smart actions can have the biggest impact. It takes a minute to forward on an e-mail to a politician or a company. A dozen such e-mails in the in-trays of that politician, or that company's head of external relations, can trigger action. The astute politician or businessperson reckons that where a dozen people take the trouble to write there may be hundreds who haven't. Maybe those dozen people have formed a small group that meets occasionally. Maybe they have become networked like the Ya-Ya Sisterhood. Take a small number of small groups, one good network, multiply the dozens of e-mails by the right number of politicians, and you have change. Take just one proactive company hit by a network with a good message at the right time in its history, and you have change. After all, these companies are made up of individuals too, and many of them might even love their kids and wish them a sane future as much as social activists do.

Then there is our money. As I have made clear earlier in the book, one of the biggest things that has to change if we are to fashion a survivable future is the way the financial sector invests. Right now billions flow out of the insurance companies, pension funds and banks into the coffers of the suicide technologies with little thought. And yet it is very often *our* money that the investors are deploying: pension contributions, banking deposits, insurance premiums – they all come from our wallets and purses. Somehow we have to correct this dysfunctional aspect of how capitalism operates. It is difficult to imagine more fertile ground for mavens. Take Paul Dickinson, founder of the Carbon Disclosure Project. He wrote to dozens of insurance and pension companies asking them to do a simple thing together: write to the big companies in which they invest, and ask them to disclose how much exposure they have to carbon, in the event that the world got serious about dealing with global warming. A simple, reasonable request, entirely consistent with the insurers' and pension funds' fiduciary responsibility to their own investors. But how wonderfully subversive. The financial institutions who sign on to the Carbon Disclosure Project letters, which go annually to the world's biggest companies, today have $2 trillion of capital behind them. Even Exxon has had to comply with their request. And so a great big sticky message goes out to the business world: oil, gas and coal mean investment risk.

When it comes to personal investments, if you have the money for such things, enlightened self-interest must extend to the best possible protection of personal assets ahead of the oil topping point. This too can help promote the chance of positive outcomes after the event. Since 1973, sharp rises in price have led to recession/stagflation, inflationary economies, occasional deflationary scares and plummeting stocks. Reviewing this history in detail, can trigger points be detected? Wall Street investor Stephen Leeb maintains that when oil prices are 80 percent or more higher than they were one calendar year earlier, the risk to stocks is overwhelming, and that this is the time at which most personal investment should be switched into deflationary hedges such as Treasury bills and bonds. If stocks are kept, he recommends that they include alternative-energy companies, blue-chip giants taking a genuine interest in alternative energy, and companies dealing in commodities vital to alternative energy, such as platinum-group metals.[330] Accepting this logic, investment in the spirit of enlightened self-interest becomes a campaigning tool, because if enough of us follow Leeb's recommended path the companies dealing with the survival technologies are strengthened ahead of the general market fall, and fare relatively better during and after it.

Many individuals who encounter the oil-depletion debate and become convinced by the arguments for an early topping point will hardly sit back and do nothing, whether they own stocks or not. An economic depression is going to hurt on so many levels. Forewarned is forearmed, and many who can will be looking already to procure the micropower and storage technologies that can give them warmth and lighting when others may well be going without, should things get that bad (and, given the state of the national grids, you can be pretty sure they will). There lie further opportunities for sticky messages and the power of small groups. They begin with the opportunity for co-procurement of solarization kit, bringing down cost, but extend far beyond that. Tell your local energy company, when they say it is too early for stand-alone systems, why you dispute that. Ask them to meet your needs. Use groups and networks to inject that message. Target the front-running energy company, environmentally, and focus on levering change in that one.

Beyond action from the home, individuals have professions. There is plenty of scope here for chipping in to the tipping point for solarization. Companies are ultimately collections of individuals trying to achieve a particular mission in a particular culture. In most companies, individuals have to be listened to. I think here of BP and its ground-breaking decision to quit the Global Climate Coalition. The main reason this was done was because the company's employees were so ashamed at what was being done and said in their name about climate change.[331]

At the community level, we need to keep the example of Woking in mind. If a small town in Surrey, England, can cut its carbon-dioxide emissions by 77 percent while making heat and electricity available to its citizens on private wires for less on the monthly bill charged by traditional energy companies, then why not any community? What a wonderful basis for collective action at the community level, whether that community is a block of flats, a church, a housing association, a village, one street in a town, or a small town itself. There are many other possibilities for community action. Car pooling can save emissions and money. Pooling of carbon credits from renewable micropower technologies such as solar electric roofs can add revenue streams to individual and community alike, while obviating the hassle for most people. It would be easy for a housing association or a church to insist on collective investment of pensions in survival technologies, for instance, or to insist collectively that all investments be socially screened, and to encourage like-minded communities to do the same.

This is not an exhaustive manifesto. If they accept the case I have made in this book about the conflation of oil depletion and global warming, imaginative individuals will be able to draw up long lists of actions holding the potential to contribute to a tipping point for solarization. I don't know the precise recipe for arriving at the tipping point. The trigger – the germ that starts the epidemic – could be in any or a combination of the things that people can do to make a difference. But I do know this. We have to find that recipe. And I have faith that, out of the ashes of the great global energy crisis, we will be able to.

EPILOGUE

The Future of the Blue Pearl

The Thinkers drifted on as Big Oversights One and Two built up to their crisis points. They had never been very good at spotting slow-burning high-consequence threats at the best of times. This time, they became distracted by one of their periodic Cycles of Hate, which made it even more difficult for them to experience a Great Awakening ahead of the crisis point. Here's how it worked. Hundreds of millions of the Thinkers believed that a man born in one part of the Middle East region of the Blue Pearl was the son of their version of God. Around a billion believed that a man born in another part of the Middle East was the prophet of their version of God. The two men were known, respectively, as Christ and Mohammed. Both were extraordinary Thinkers who lived amazing lives, but they were both let down badly by some of their followers after their deaths. The followers of the two men had rather profound differences of opinion, not just about the nature of their Gods, but also about stuff like how to dress. By and large they lived and let live, did a lot of trading in the long tradition of the Thinkers, and often became friends despite all the arguments about the length of skirts. But there was a big problem. Most of the oil lay under the lands of the followers of

Mohammed while the Number One Consumer was run by the followers of Christ.

The Number One Consumer would stop at nothing to access the oil in the Number One Producer. Not to mention Numbers Two to Twenty, for that matter. Mostly they did this by trading, but they also liked to jump in their jet aircraft and drop a few bombs, given half a chance. The Number One Consumer had the biggest military the Thinkers had ever assembled. They had thousands of bases in hundreds of countries. In terms of unmatchable firepower and geographic reach, there had never been an empire like the Number One Consumer Empire in all the history of the Blue Pearl. And the Empire liked to, you know, *use* some their kit from time to time.

THE DAY THAT CHANGED EVERYTHING

Here's how this particular Cycle of Hate began. The problem was that both sets of followers had their zealous factions, the Fundamentalists. A good few Fundamentalist Mohammedans took great exception to the Consumer Empire, especially the foreign bases part of it. They wanted the Empire to quit its bases in their holy lands and go back to its own territory. They couldn't hope to use the power of oral persuasion. Neither could they think of taking the Empire on in open battle. So they did what dissident Thinkers with a stomach for violence had done for centuries; they turned to terrorism. For a fair while, all they could do was blow up overseas outposts of the less protected parts of the Consumer Empire. The Empire, far away in its homeland, was complacent. Nobody had ever attacked it at home. Some Thinkers from a small group of islands with a strong warrior tradition and an insane machismo had once attacked one of the Empire's offshore islands. They lived to regret it, big time. Anyway, the Empire's complacency was misplaced. The

Fundamentalist Mohammedans who had turned to terrorism worked out a way to fly hundreds of litres of jet fuel at 250 metres per second into iconic buildings in the heart of the Number One Consumer's homeland. That sure was a Day That Changed Things.

This awful murderous act could not have come at a worse time, from the perspective of the Blue Pearl herself or the many hundreds of millions of Thinkers around the planet already watching the antics of the Consumer Empire aghast. The reason was that Fundamentalists of the Christian type were in control of the Consumer Empire at the time. This was just the recipe for a Cycle of Hate. Did they ask any questions about the root causes of why maniacs with degrees in engineering had done this horrible thing with the passenger jets? Not a bit of it. Was there any discussion about the long history of nation states imperilling their own security by undermining the security of other nation states? None at all.

THE EMPIRE STRIKES BACK ... AND BACK ... AND BACK

The Consumer Empire leapt into its bombers. It began by carpet bombing a Middle Eastern nation state where some not-very-thoughtful Thinkers had let the terrorists train their cadres. Liberal Thinkers all around the Blue Pearl watched their TVs in horror as supposedly smart bombs took out mud-huts and unidentified vehicles on desert roads, but also hospitals, homes and mine-clearance staff from neutral nation states trying to help clear up after an earlier contretemps in the region. With each detonation, new Mohammedans joined the queues at the terrorists' recruiting offices. The Consumer Empire was demonstrating a very old lesson from the history of the Blue Pearl. The best way to make enemies, swell their ranks, stiffen their resolve and turn them to terrorism is to bomb their mothers and sisters.

Not content with the carpet bombing, awesomely effective as it had been, the Consumer Empire decided to invade a second Middle Eastern nation state. This one was Oil Producer Number Fifteen. The reason given by the Empire for the invasion was that Number Fifteen was run by a tyrant who didn't hesitate to use deadly force against his own people, and the Empire wanted to bring democracy to the people (after they had Shocked and Awed them with a real-life demonstration of firepower, that is). That was a strange argument, because Producers Number One, Two, Four, Six, Ten, Twelve, Thirteen and Fourteen were all run by varying degrees of tyrant who opposed democracy and didn't hesitate to use deadly force on their own people.[332] There wasn't any talk of invading *them* just yet. And anyway, quite a few of the other Producers, although democracies, knew a thing or two themselves about the use of deadly force on their own people.

Liberal Thinkers were quick to point out that Producer Number Fifteen was also Oil Reserves' Holder Number Three. Could it just be that Consumer Number One had just a little interest in this fact? Could it be that they were stoking the state-of-fear and desire-for-revenge factors in their homeland as a smokescreen for a little expeditious oil grabbing?

Whatever, the invasion went ahead. The defending army was duly Shocked, Awed and Generally Incinerated. The smart bombs took out buildings surgically, as they were supposed to do, but unfortunately some of them proved to full of Thinkers who were not obviously of a military persuasion. A Mohammedan TV station filmed all this and broadcast the pictures around a generally appalled planet. Consumer Number One bombed their building too.[333] The Cycle of Hate deepened predictably. The invaders added atrocities of their own to those of the terrorists. They liked to use their digital cameras to record them, and so Picture Nasties inevitably found their way on to TV screens and into newspapers. Incensed, Mohammedans took up arms to combat the invading Christians in ever greater numbers, so much so that the Consumer Empire found difficulty classifying them all as terrorists. But of course, many of these newly livid Thinkers were quite prepared to dish out a bit of terror along with the bullets. The terror masters rejoiced in the success of their recruiting plan, marvelled at how the Consumer Empire had played along with their every scripted move, and dreamed of weapons even more terror-worthy than fuel-laden passenger jets.

In both the Consumer Empire and the Middle Eastern nation states, Fundamentalism thrived amid all this mayhem. In the Consumer Empire, many Thinkers held a belief that their version of God would let them trash the Blue Pearl and still join Him in His Heaven. Indeed, the more quickly they trashed it, the more quickly they would get to enjoy what they thought of as "the rapture". Rapturist Thinkers tended to care nothing about the fuel efficiency of their horseless carriages, didn't give a damn about the Alternatives to oil, and were very keen on burning coal. They also figured that all-out war might be another way to join their version of God more quickly than letting things run their normal course, so they weren't averse to that either. They thought of it as the Third World War: The Clash of Civilizations. The Fundamentalists on the other side were no less keen for a taste of Armageddon. These were Thinkers who were so angry and confused that they tended to think they could get to their God's version of heaven by strapping a bit of semtex to their torso and suicidally taking out a few boy soldiers in a bus queue, even if old ladies and kids were standing there along with them.

For Thinkers who believed in Cosmopolitan Tolerance – simply stated, having a stab at learning from the lessons of history[334] – things were beginning to look really very bleak at this point.

BIG OVERSIGHTS ONE AND TWO SHOW UP

To add to their woes, while everyone was so distracted by the Cycle of Hate, evidence that the Blue Pearl was slowly cooking started to crop up thick and fast. Most worryingly, scientists in Consumer Number One published seven million temperature readings taken over forty years to depths of more than 2,000 feet in the oceans. They professed themselves "stunned" by the extent of the warming they found. They said it proved once and for all that greenhouse gases from oil, gas and coal burning were heating up the planet alarmingly.[335] The Fundamentalists who led the Consumer Empire, and who had strong links with oil interests – to put it mildly and politely – continued to take no notice. Fortunately for the Cosmopolitan Thinkers, at least in the long run, the Fundamentalists weren't taking much notice of the economy either. Financing all the oil profligacy and the bombers involved rather a lot of borrowing from other nation states. The Thinkers had always found the running of empires an expensive affair. This would prove to be the downfall of the Fundamentalists who ran Consumer Number One.

The biggest single user of oil on the Blue Pearl was the military of Consumer Number One. Despite all their firepower, they became bogged down in their occupation of Producer Number Fifteen. Despite all the untapped oil reserves below the surface, there were days that the angry locals were so active with their semtex and stuff that barely enough oil came out of the ground to keep the military's tanks full, much less contribute to the gas guzzlers back in the homeland. Little by little the Trading Thinkers, watching all this, and the mad scramble for oil unfolding in other parts of the Blue Pearl, came to realize that the news about the black stuff was not as they wanted. They assumed Producer Number One could up the pump rate to meet demand. It couldn't. They assumed that Producers Two to Twenty might be able squeeze more out.

They weren't. They expected a lot more oil could be found in Producers yet to join the Top Twenty. It wasn't. Worse than that, the oil companies in Consumer Number One didn't even seem to be trying that hard to find it, for some reason. They should have been spending money by the barrel-load looking. They were spending it only by the bucketful. So it was that one week – one apocalyptic week – a germ of panic took root and then spread like wildfire through the Trading Thinkers. The price of oil, on which the Thinkers had allowed the stability of their economic system to rest, began to climb towards the ceiling. As the panic spread like an epidemic the following week, the oil price went through the roof.

COLLAPSE

The crisis of Big Oversight One played out on television images around the Blue Pearl. Frantic Oil Traders screamed at each other on trading floors, eyes wild and hair akimbo. These were hardly scenes conducive to calm in other markets, and share prices began to slide. "Oil Running Out," the headlines read. It wasn't, it was merely Half Gone, and so becoming very, very expensive – very, very quickly. Leaders of the Consumer States and the Producer States got together for a crisis summit, looking appropriately grim in their suits and flowing robes, respectively. They could think of nothing much to say that would ease the panic, so it spread further. Producer Number One had allowed a mountain of consumer debt to pile up, most of it in houses. The price of houses collapsed. Stock markets crashed. Within the space of a month, the wealth of the Thinkers – little more than a pile of paper at the best of times, even with confidence about the future high among the Trading Thinkers – shrivelled. The inescapable consequences of the crisis then rolled out in slow motion. Companies went bankrupt by the hundreds and then thousands. Workers fell into unemployment by the hundreds of thousands and then millions. Once affluent cities with street cafés now had queues at soup kitchens and armies of beggars on the streets. The crime rate soared. The Blue Pearl had always been a dangerous place, but now it had become a tinderbox.

All this had happened to the Thinkers once before, in fact: 1,929 years after Christ's birth. Then the economic depression had had nothing to do with oil, just a general catastrophic collapse of confidence by Trading Thinkers. It had taken the Thinkers many years to dig themselves out of the mess. In the aftermath of the stock-market crash at that time, the problems had been compounded by the emergence of a further category of Fundamentalist Thinker, the Fascists. Fascist Thinkers believed in having one powerful leader and a big secret police force with well-kitted-out torture chambers. Democracy and Cosmopolitan Tolerance were most definitely not on the agenda of these guys, although they tended to pretend otherwise in the early stages of their meteoric rise. They fed on the anger of the newly unemployed poor. They whipped up sentiment against a third category of Religious Thinker, who followed neither Christ nor Mohammed, and believed that the son of their version of God had yet to visit the Blue Pearl. These Thinkers were often Traders, and so residually rich among all the general hardship. The Fascist Thinkers burnt their homes, and then herded them into trains and sent them to concentration camps. In the worst of the many examples of genocide in the Thinkers' history, the Fascist Thinkers then began systematically to exterminate these poor Trading Thinkers by the million. Nobody came to their rescue because by this time the Fascist Thinkers had plunged the Blue Pearl into a planet-wide war.

Now, 2,010 years after Christ's birth, the whole bloody business looked as though it might be beginning all over again. Fascists once more crawled out of the woodwork and got to work on the poor. This happened in many nation states. The problems this time around were even worse, for two reasons. First, in the wake of the incident with the passenger jets and the iconic buildings, Consumer Number One had fanned a state of fear among its populace. In that environment, it had put in place many State Instruments of Repression. Emergency laws permitted incarceration without trial. Special prison camps had been set up. Laws on torture had been relaxed. Identity cards had been introduced. Fundamentalist judges had been appointed to the supreme judiciary to enforce the laws. The Fundamentalist Thinkers who controlled Consumer Number One at the time of the invasion of Producer Number Fifteen had sort of favoured democracy themselves, if not being averse to a little rigging of it. Giving them the benefit of the doubt – if for no other reason than that they seemed so keen on democracy as a reason for invading Producer Number Fifteen – they probably didn't fully realize how easy they were making it for Fascists to rule, if the latter could just but get themselves into power.

The second reason the problems were worse was because Big Oversight Two now began to make its presence felt around the Blue Pearl with a vengeance. Thinkers who had been impoverished virtually overnight now watched aghast as their food and water supplies dwindled in the face of a climate seemingly going awry. In Consumer Number One, appalling and prolonged droughts spread through the south and middle of the nation state, decimating the grain and other harvests. At sea, fish catches fell off a cliff. The Blue Pearl's warming ocean water was tending to stratify and acidify, so nutrient levels were falling, meaning that phytoplankton stocks were plunging, meaning that fish stocks were collapsing. All of a sudden, protein was becoming a luxury. As for the water, there was so little to spread around that the Thinkers even had to stop using it on their golf courses.

The Fascists fed on all this and whipped up a pile of support right around the Blue Pearl. In a few nation states, they actually got themselves elected. In Consumer Number One, Cosmopolitan Thinkers ousted the Fundamentalist Thinkers who had launched the disastrous invasion of Producer Number Fifteen, overseen a catastrophe in the economy, and failed to protect those who had elected them from Big Oversight One. But the Cosmos fared poorly. Who wouldn't, given the enormity of the train wreck the Fundis had bequeathed them? Amid general and escalating civil unrest, the Fascists in Consumer Number One decided to go down the road of armed insurrection – a coup – just as they had seventy years before.[336] Behind closed doors, a leading politician, a leading general and several well-tooled billionaires got planning. Around the nation, cadres of well-armed, well-organized and extremely unpleasant Thinkers secretly urged them on.

Energy supply was a nightmare, almost everywhere. Some oil was available for the Thinkers' horseless carriages, but at huge expense, and only after queuing for two days. Gas supplies, now assumed to be heading for a topping point of their own, were also astronomical. Power prices were sky high in consequence. Years of underinvestment in an aged power grid now took their toll. Rolling power cuts spread across the land. Coal was being scrabbled out of the ground apace. But one thing the new Cosmo leaders of Consumer Number One were able to do, amid all their trials and disappointments, was ride a wave of concern about the spreading droughts and the dying fish stocks, and unlock a well of exploitation of the Alternatives.

RENAISSANCE

And it was here – amid economic depression, dreadful suffering across the planet and a rising tide of authoritarian horror – that the seed of hope was finally planted. The Thinkers began to discover, at first in relatively small numbers, that it wasn't that difficult to turn their backs on oil, gas and coal, and everything that went with them. Amid the hard times, many Thinkers developed an understandable interest in building their own homes. It proved relatively easy to kit these out with Alternative-Energy gadgets that combined electricity and heat derived from the sun. With this lot, Thinkers could meet energy demand in the home at any time, day or night, with lots of electricity left over to make hydrogen from water, for storage in the home and use in their horseless carriages. Just a few short years before, utility companies had owned the power plants, oil companies had owned the filling stations, and both had tended to price-gouge Consumer Thinkers ruthlessly given half a chance. Now individual Thinkers were able to act as utility and fuel companies themselves, their lights never went off, and there were no fuel shortages, at least locally. Quite the reverse, in fact. The new generation of homes could generate much more power than they needed, and so export it to the electricity grid, local or national. What remained of the construction industry noticed this trend, and copied it. Before you could say "Hey, this is a tipping point", very few buildings were going up anywhere that weren't temples to Alternative Energy.

The Alternative-Energy homes proved popular like few products ever had before in all the history of Thinkers' long history of Consumerism. The Alternatives began to be fitted to existing homes, and to bigger buildings. The phenomenon went beyond energy. It seemed to be engendering a new interest in the very idea of community. Seeking security beyond keeping the lights on and the horseless carriages fuelled up, Thinkers relocated to linked networks of these homes, looking to share other security-building measures. The new Community Thinkers plugged their horseless carriages into the local electricity grid while parked, feeding surplus electricity in, generating income on the side, and enough energy to begin local economic activities. Local manufacturing of Alternative-Energy kit sprouted everywhere. All manner of locally produced goods for local needs came alongside this new feature of Thinker behaviour. The Community Thinkers organized trading pools to trade green electricity for the best prices they could find in the carbon markets set up by the Cosmo governments in their efforts to head off the worst excesses of Big Oversight Two. They organized pension clubs, and used money made from Alternative-Energy generation to top up their funds. Growing numbers of Thinkers seemed to have learned the lesson about interdependence. The Community Thinkers lived and worked quite happily in their neighbourhoods, but they shared ideas and plans and communicated their excitement with other like-minded communities in their own and other nation states. They wanted to know that their neighbours were as secure as they were!

As the craze for Alternatives spread, so dependence on oil first withered and then vanished. The missile bill in the Consumer Empire dropped to a fraction of what it once was, which is just as well, because nobody could afford it any more anyway. The troops returned to their homeland. In fact, not to put too fine a point on it, the Consumer Empire began to look less and less like an empire with every passing day. With the unique clarity that comes with hindsight, Historian Thinkers expressed amazement that those oil wars had cost over a billion dollars a day. Where did that feature in the pricing of a barrel of oil? they asked. And those Fundis had had the gall to say they believed in market forces!

The spread of Alternatives in the poorer nation states began to erode the equity gap, narrowing the breeding grounds of hatred of the shrinking Consumer Empire and its lackeys. A famous Thinker in Oil Producer Number Eleven said in 2000 that "the battleground of the twenty-first century will pit fundamentalism against cosmopolitan tolerance".[337] All of a sudden, the Cosmos – feeding on the advent of the Communities – looked to be in with a chance of seeing off the Fundis and the Fascists. Their rallying cry became local energy for local economies for local community, or words to that effect. Slick marketing slogans didn't have the cachet they once had, you understand.

Fresh-water supplies were dwindling fast in the face of pollution, unsustainable mining of aquifers, and the droughts which by then were striking regularly. But as the Alternatives proliferated, many coastal towns and cities developed their own solar desalination plants. Suddenly fresh water was available again. Agriculture became possible in the hinterland of the big desalination plants in arid regions such as the Middle East. This contributed further to the slow erosion of the terrorism epidemic. A form of peace finally descended on the Middle East. People were too busy farming to bother about their semtex and Kalashnikovs.

Descriptions of the air quality in many cities at the time of the Great Awakening to Big Oversight One beggared belief. Estimates of the cancer and asthma tolls among Thinkers from breathing partially combusted hydrocarbons were a horror story – though of course the Thinkers were in deep denial about that at the time too. This horror began to go away. Many cities had hardly any particulates in their air. Hospitals slowly became less stressful places, and the standard of the food even improved.

Many of the routine assumptions current among Trading Thinkers in the Age of Oil Addiction simply evaporated. For example, once it had made sense that widgets should be manufactured on one side of the Blue Pearl because labour was cheap (and often underage) and shipped to markets on the other side, at great but entirely unaccounted pollution cost. Thinkers began to talk as much about "localization" as they did globalization.[338] They still had globalization of communications of course. That was a big factor in the renaissance of hope after the Second Great Depression.

Despite all the focus on the local, the Cosmos also presided over a resurgence of public involvement at the national and global levels. Slowly the early success of the Fascists ebbed. The coup plotters in Consumer Number One were exposed. Before the Great Awakening, even though people expressed huge concerns about the state of the Blue Pearl in opinion polls, they had stayed away from elections in growing numbers in many Nation States, notwithstanding the fact that democracy itself was spreading. There was widespread distrust of politicians. After the changes, the culture of independence and localization – much helped by the transparency made possible by the continuing information revolution – finally meant not just horizontal but vertical spread of democracy. Thinkers began to vote routinely on things they wouldn't have dreamt of in the time of the Great Oil Addiction. They did it simply by logging on to the internet at specific times. They had, in a very real sense, "democratized democracy".

The Thinkers by now began to look as though they might even be on the way to stabilizing their global population. The key, of course, was the education and consequent emancipation of their females, as was clear even before Big Oversight One in the thirty-one nation states that had by then succeeded in stabilizing their populations.[339] The spread of Alternatives in the poorer nation states, and the delivery of electricity to the third of the Thinkers who hadn't had it at all before the Great Awakening meant a very simple thing: light at night everywhere. Enter the textbooks.

Slowly it also began to look as though the worst excesses of Big

Oversight Two could be avoided. The infrastructure challenges associated with adapting to the delayed effects, especially of the slowly rising seas, remained huge. The slow-burn impacts, such as the devastation of coral reefs, remained sickening to behold. But concentrations of the heat-trapping gases were beginning to level off in the atmosphere. Scientists were becoming less anxious. Innovations in the technologies of both abatement and adaptation were breaking out all over in the Alternative Communities.

And as the Thinkers looked back, with their memories of history dimming as they always seemed to do, it seemed oh so difficult for them to understand why they had taken so long to see, and do, all these obvious things.

NOTES

1. For further information here, see the Geological Museum's exhibition on the origins of life on Earth.

2. The volcanic hypothesis for the mass extinction at the end of the Permian period involves a basalt eruption in Siberia that spewed a vast flood of lava along a split in the Earth's crust. The lava flow covered 200,000 square kilometres. The eruptions lasted millions of years, though the extinctions seemed to have happened in less than 100,000 years. Massive ash clouds would have caused global cooling. When they cleared, temperatures could have risen as much as 5°C on pre-eruption levels because of carbon dioxide emitted by the volcanoes. Advocates of a meteorite argue that a rock 15 kilometres across could have had the same effect. They point to meteorite fragments in sediments of the time as evidence. (*The Day the Earth Nearly Died*, BBC Horizon, 5 December 2002.)

3. *Homo erectus*, the precursor to modern humans, evolved in Africa between 1.8 and 1.9 million years ago. Around 230,000 years ago the Neanderthals appeared. These were the first hominids to have the same brain size as modern humans. They were replaced by *Homo sapiens*, who originated around 200,000 years ago and began their global spread out of Africa around 120,000 years ago. For further information, see James F. Luhr, ed., *Earth*, Dorling Kindersley, 2004.

4. Edward Lloyd opened a coffee house in 1688, establishing a clientele of ships' captains, merchants and ship owners and a reputation for trustworthy shipping news. Lloyd's Coffee House became the place for obtaining marine insurance. In 1774 the modern Lloyd's of London opened. In 1871 Lloyd's was incorporated by an Act of Parliament, giving it formal legal basis. (For

more on Lloyd's see www.lloyds.com.)

5. Thomas Edison demonstrated electricity in 1882.

6. The first oil was drilled in 1859, in Pennsylvania.

7. Henry Ford built his first car in 1896.

8. 8.18 billion tonnes of oil equivalent. (International Energy Agency [henceforth IEA], *Key World Energy Statistics – 2004 Edition*, 2004.)

9. The Framework Convention on Climate Change was signed by heads of state in Rio de Janeiro at the Earth Summit in 1992. The Convention came into force on 21 March 1994. As of 24 May 2004 the Convention had received 189 instruments of ratification. (http://unfccc.int/essential_background/convention/status_of_ratification/items/2631.php.)

10. See my book *The Carbon War*, Penguin, 2000, for a detailed historical justification of this strong statement.

11. *The Carbon War*, quoted at note 10, has an account of the emergence of the international global-warming threat assessment in the years 1988 to 2000, at least in so far as it became perceived by most governments at the time the Kyoto Protocol to the Convention on Climate Change was signed in 1997.

12. A recent exhaustive study suggests that up to 35 percent of species will be "committed to extinction" by 2050 if the global-warming commitment involves an average global temperature of more than 2°C above the pre-industrial average. (Chris Thomas et al., "Extinction risk from climate change", *Nature*, 8 January 2004.) For a wider summary of the problem, see the UN's "Millennium Ecosystem Assessment Report", 2005.

13. BP's latest advertising, carried on a proliferating number of London taxis at the time of writing, boasts about their Ultimate Fuel.

14. The Hummer. A clip in a Rising Tides Campaign video shows Jeremy Clarkson informing the audience of one of his TV programmes that he is driving one of these huge SUVs at one mile per gallon. He actually finds it funny.

15. Chris Skrebowski, "Joining the dots", presentation to Energy Institute conference "Oil depletion: no problem, concern or crisis?", London, 10 November 2004.

16. A barrel of oil contains 42 US gallons, weighs 0.1364 tonnes and can fill a typical gas tank several times. The six barrels comes from "The price of steak", *National Geographic*, June 2004, p. 98. The article cites a 1,250 lb steer requiring 283 gallons. 1 barrel = 42 US gallons, 6 barrels = 252 gallons @ 30 miles per gallon = 7,590 miles. Distance between New York and LA is 2,800 miles.

17. US Energy Information Administration (www.eia.doe.gov).

18. In the IEA's *World Energy Outlook 2004* reference case, world oil demand

increases by 1.6 percent annually to 121 million barrels per day in 2030 (www.iea.org).

19. "Energy in focus: BP statistical review of world energy", June 2004. Available as a pdf from www.bp.com.

20. http://www.eia.doe.gov/emeu/cabs/pgulf.html, "Country analysis brief: Persian Gulf", September 2004. In 2003, about 90 percent of oil exported from the Persian Gulf transited by tanker through the Straits of Hormuz at a rate of 15.0–15.5 million barrels per day. It goes eastwards to Asia (especially Japan, China and India) and westwards to Western Europe and the United States.

21. Paul Roberts, *The End of Oil: The Decline of the Petroleum Economy and the Rise of a New Energy Order*, Bloomsbury, 2004, citing energy-efficiency guru Amory Lovins.

22. http://www.fueleconomy.gov/feg/FEG2005GasolineVehicles.pdf, "2005 model year vehicles". The Prius does 60 miles per gallon for city driving, 51 miles per gallon for highway. It went on sale in Japan on 10 December 1997, and was launched in US August 2000. (http://www.motortrend.com/roadtests /alternative/112_news46/, "Is Toyota Prius the most important 2004 model?") I have tended throughout the book to use imperial measurements as well as the US vocabulary autos and gasoline.

23. Michael Klare, *Blood and Oil: The Dangers and Consequences of America's Growing Dependency on Imported Petroleum*, Metropolitan Books, 2004, p. 46.

24. The second Iraq War had cost more than $140 billion by August 2004, according to the Center for American Progress. The Congressional Budget Office estimates costs under three scenarios for occupation of Iraq, ranging from $179 billion to $392 billion over the period 2005–2014. (Letter from [non-partisan] Congressional Budget Office to Senator Kent Conrad, ranking member of the Committee on the Budget, http://usgovinfo.about.com/gi/ dynamic/offsite.htm?site=http://www.cbo.gov/.)

25. Elizabeth Economy, *The River Runs Black: The Environmental Challenge to China's Future*, Cornell University Press, 2004.

26. Association for the Study of Peak Oil and Gas (henceforth ASPO) newsletter, December 2004.

27. Different types of oil have different prices. The two most commonly used prices, and the two I use, are Brent Crude and West Texas Intermediate. The prices of these are close, and I have not distinguished between them in the book.

28. At the time of writing, the oil price had passed $62 per barrel.

29. www.wikipedia.com.

30. For good accounts of the Depression see Studs Terkel, *Hard Times: An Oral History of the Great Depression*, The New Press, 2005; Harold James, *The End*

of Globalization: Lessons from the Great Depression, Harvard University Press, 2002.

31. This exchange, with John Schiller, an emissions planning associate with Ford's Emissions Control Planning staff in Michigan, is described on pp. 173–174 of *The Carbon War*, quoted at note 10.

32. I learned later, taking my daughter around a zoo, that the geographic range of the salt-water crocodile extends all the way up South-East Asia into Baluchistan.

33. John Platt and Jeremy Leggett, "Stratal extension in thrust footwalls, Makran accretionary prism: implications for thrust tectonics and section balancing", *Bulletin of the American Association of Petroleum Geologists*, vol. 70, pp. 191–203.

34. These mudstones were younger than the more recent of the two major phases of oil formation in the planet's history (the two Great Underground Cook-Ups in The Story of the Blue Pearl), so our oil would have been unusual. But the fact that we could see smears of oil bubbling from the ground in places showed us that younger oil had been formed.

35. Jeremy Leggett and John Platt, "Structural features of the Makran fore-arc from LANDSAT data", in B. Ul Haq and J. Milliman (eds), "Marine geology and oceanography of Arabian Sea and coastal Pakistan", National Institution of Oceanography Special Publication, pp. 33–43.

36. Kenneth Deffeyes, *Hubbert's Peak: The Impending World Oil Shortage*, Princeton University Press, 2001, p. 7.

37. *Hubbert's Peak*, quoted at note 36.

38. Svend Duggen, Kaj Hoernle, Paul Van Den Bogaard, Lars Rupke and Jason Phipps Morgan, "Deep roots of the Messinian salinity crisis", *Nature*, 10 April 2003. The desiccation of the Mediterranean Sea between 5.96 and 5.33 million years ago was one of the most dramatic events on Earth during the Cenozoic era (the most recent 65 million years, or the time after the dinosaurs). It resulted from the closure of marine gateways between the Atlantic Ocean and the Mediterranean Sea, the causes of which remain little understood but may well best be explained by uplift of Iberia and North Africa as a result of tectonic plate movements: the African plate pushing north against Europe.

39. For a more detailed account of reservoirs and traps, see Chapter 3 of *Hubbert's Peak*, quoted at note 36.

40. For a more detailed account of the techniques of oil exploration, see Chapter 4 of *Hubbert's Peak*, quoted at note 36.

41. IEA, *World Energy Outlook 2004*, p. 96, Fig. 3.13: Non-Conventional Oil Resources Initially in Place.

42. For a more detailed account of drilling methods, see Chapter 5 of *Hubbert's*

Peak, quoted at note 36.

43. For further detail see Roger Anderson, "Oil production in the 21st century: recent innovations in underground imaging, steerable drilling and deepwater oil production could recover more of what lies below", *Scientific American*, March 1998, pp. 68–73.

44. Gordon Cope, "Synfuel excess", *Petroleum Review*, November 2004.

45. "The next scandal? In the wake of the Shell scandal, there are doubts about reserves at other firms", *Economist*, 13 November 2004.

46. "Energy in focus", quoted at note 19.

47. Reports from 2000 to 2003 cite the source of data as follows: "With the exception of Azerbaijan and Kazakhstan, the estimates (of proved reserves) are those published by the *Oil & Gas Journal*, plus an estimate of natural gas liquids for USA and Canada. Reserves of shale oil and oil sands are not included." Canadian oil sands "under active development" were included in proved reserves for the first time in 2004.

48. Bob Williams, "Debate over peak-oil issue boiling over, with major implications for industry, society", *Oil & Gas Journal*, vol. 101, no. 27, 14 July 2003, pp. 18–29, quoting Jean Laherrere.

49. C.J. Campbell, "The essence of oil depletion", *Multi-Science*, 2003; Powerpoint presentations in CD accompanying the summary booklet C.J. Campbell, "The truth about oil and the looming energy crisis", available from www.aspo.org.

50. ASPO newsletter, October 2004.

51. Julian Darley, "A tale of two planets", a report on the conference "Future of global oil supply: Saudi Arabia", held at CSIS, Washington DC, 24 February 2004 (*From The Wilderness* special).

52. Dmetri Savastopulo and Carola Hoyos, "Saudi Aramco dismisses claims over problems meeting rising global demands for oil", *Financial Times*, 27 February 2004.

53. All information in this paragraph from "Big oil's biggest monster: Saudi Arabia's Aramco, the biggest and most powerful oil firm, has revealed some of its secrets", *Economist*, 8 January 2004.

54. "Debate over peak-oil issue boiling over", quoted at note 48.

55. Mamdouh Salameh, "How realistic are Opec's proven oil reserves?", *Petroleum Review*, August 2004.

56. Sadad al-Husseini, "Rebutting the critics: Saudi Arabia's oil reserves production practices ensure its cornerstone role in future oil supply", *Oil & Gas Journal*, 17 May 2004. Despite the title, al-Husseini defends Saudi Arabia's 260 billion barrels of reported proved reserves as follows: "Given the fact that the discovered but underdeveloped Saudi reservoirs make up about

130 billion barrels of the kingdom's total reserves, the addition of new proven reserves through future reservoir developments is a foregone conclusion." In this paper, he also warns of the dangers of high-risk production practices of the kind Matthew Simmons thinks have already harmed the large reservoirs (see note 58).

57. ASPO newsletter, December 2004.

58. "Expert: Saudi oil may have peaked", al-Jazeera Economy News, 20 February 2005.

59. These events are described by Kenneth Deffeyes, a contemporary of Hubbert's at the time, in *Hubbert's Peak*, quoted at note 36.

60. Stephen Goodwin, "Hubbert's Curve", *Country Journal*, November 1980, pp. 56–61. This paper, written with the benefit of a quarter of a century's hindsight, recounts the sequence of Hubbert's arguments in 1956, and the aftermath.

61. "Hubbert's Curve", quoted at note 60.

62. "Hubbert's Curve", quoted at note 60.

63. Statistics in this paragraph are from "The truth about oil", quoted at note 49. Table 1 of that document summarizes all countries and regions.

64. "The truth about oil", quoted at note 49. The Exxon paper quoted is by Harry J. Longwell, "The future of the oil and gas industry: past approaches, new challenges", *World Energy*, vol. 5, no. 3, pp. 100–4.

65. ASPO plots the crossover year as 1981; Ken Chew of the consultancy IHS professes that it is 1986, if oilfields are corrected for enhanced recovery (Ken Chew, "Oil depletion – the database", Energy Institute conference quoted at note 15).

66. Francis Harper, in panel discussion during the Energy Institute conference quoted at note 15.

67. Data in this paragraph from Colin Campbell's ASPO database.

68. *Hubbert's Peak*, quoted at note 36.

69. Chris Skrebowksi, presentation to a seminar for parliamentarians, London, 6 July 2004.

70. Francis Harper, "Oil peak – a geologist's view", Energy Institute conference quoted at note 15. The first statement comes from his written abstract, the second from discussion.

71. The 2004 Fortune 500 list shows that BP has overtaken ExxonMobil as the world's largest oil company, with a sales volume of US$232.5 billion.

72. The early toppers estimate future discovery from "creaming curves", which plot cumulative discovery over time, or against the number of "wildcat" wells drilled in a region. Such curves consistently tail off over time in different provinces and countries.

73. ASPO newsletter, October 2004.

74. Roger Bentley, "Global oil depletion: viewpoints in collision", Powerpoint presentation at Energy Institute conference quoted at note 15.

75. "US Geological Survey world petroleum assessment 2000 – description and results", US Department of the Interior, US Geological Survey, http://pubs.usgs.gov/dds/dds-060/index.html#TOP. The full report, which includes deep-water oil, is 32,000 pages long.

76. "Exxon Unleashed: How the world's most powerful corporation plans to dominate the new age of oil exploration", *Business Week*, 9 April 2001; "Black Gold Rush – BP Amoco: the hottest prospect in the oil patch", *Forbes*, 2 April 2001. Exxon was gearing up to expand its oil and gas output for the first time since the 1970s. The $232.7 billion revenues Exxon had returned the previous year made it the number one oil giant in 2001, well ahead of Shell's $191 billion and BP's $148 billion. Both Exxon and BP planned to hike their output by 3 percent, in line with industry projections of growth in the global oil market. To do this and replace shrinking output in mature fields, Exxon planned to invest $10 billion in exploration and production in 2001 alone.

77. This didn't happen. 2004 production from water more than 500 metres (1,600 feet) deep was about 10 percent. Source: Francis Harper, personal communication.

78. Ivan Sandrea, "Deep-water oil discovery rate may have peaked; production peak may follow in 10 years", *Oil & Gas Journal*, vol. 102, no. 28, 26 July 2004, p. 18.

79. "Oil peak – a geologist's view", abstract quoted at note 70.

80. Colin Campbell, personal communication, March 2005.

81. "Deep-water oil discovery rate may have peaked", quoted at note 78.

82. Information in this section from Bob Williams, "Progress in IOR technology, economics deemed critical to staving off world's oil peak production", *Oil & Gas Journal*, vol. 101, no. 30, 4 August 2004, pp. 18–25.

83. "Oil peak – a geologist's view", discussion quoted at note 70.

84. Jeffrey Currie et al., "The sustainability of higher oil prices: revenge of the Old Economy, part 2", Goldman Sachs Global Commodity Research, 8 June 2004.

85. "Joining the dots", presentation, and comments in discussion, quoted at note 15.

86. *Hubbert's Peak*, quoted at note 36, p. 10.

87. Information in this section from "Progress in IOR technology", quoted at note 82.

88. Matthew Simmons, "Energy: a global overview", seminar for Deloitte and Touche, available from the website of Simmons and Co,

http://www.simmonscointl.com/files/Deloitte%20&%20Touche.pdf.

89. This section based on Bob Williams, "Heavy hydrocarbons playing key role in peak-oil debate, future energy supply", *Oil & Gas Journal*, vol. 101, no. 29, 28 July 2003, pp. 20–27; IEA, *World Energy Outlook 2004*, p. 114: Non-Conventional Oil Production Prospects. According to the latter, total oil-shale resource is 2.66 trillion barrels (38 percent of 7 trillion), the total extra heavy (crude) oil is 1.61 trillion barrels, and tar sands and bitumen are 2.73 trillion barrels.

90. "Canadian crude oil forecast", Canadian Association of Petroleum Producers press release, 15 July 2004. Investment spending related to oil-sands development is projected to exceed C$30 billion over the next decade as new projects or expansions to existing oil-sands operations are built.

91. "Heavy hydrocarbons", quoted at note 89, and IEA, *World Energy Outlook 2004*. Total non-conventional production is projected to increase from 1.6 million barrels per day in 2002 to 3.8 million barrels per day in 2010 and 10.1 million barrels per day by 2030: this last figure representing 8 percent of projected global oil supply.

92. "Synfuel excess", quoted at note 44.

93. "Alberta industry receives warning on water usage", *Business Edge*, 24 June 2004.

94. "Synfuel excess", quoted at note 44.

95. National Energy Board (Canada), Annual Report 2003, http://www.nebone.gc.ca/Publications/AnnualReports/2003/Annual Report2003_e.pdf.

96. This section based primarily on "Heavy hydrocarbons", quoted at note 89, except one tonne per half-barrel, and projected contribution by 2012, which come from Bob Holmes and Nicola Jones, "Can heavy oil avert an energy crisis?", *New Scientist*, 2 August 2003.

97. Peter Fairley, "Digging a carbon hole for Canada: will oil sands projects be white elephants in the post-Kyoto world?", *Alberta Views*, http://www.albertaviews.ab.ca/marapr03/marapr03carbon.pdf. Cites Natural Resources Canada figures showing that the process of converting oil sand to synthetic light crude takes 104 kilograms of carbon dioxide per barrel in 2000, down from 141 kilograms in 1990, compared to 30 kilograms for conventional crude. CO_2 reduction per barrel of synthetic crude is expected to "bottom out" at 90 kilograms in 2005.

98. "Oil peak – a geologist's view", discussion quoted at note 70.

99. "Synfuel excess", quoted at note 44.

100. Remarks by Lee R. Raymond, Chairman and CEO, ExxonMobil Corporation, OPEC international seminar, Vienna, Austria, 16 September

2004 (http://www.exxonmobil.com/Corporate/Newsroom/SpchsIntvws/ Corp_NR_ SpchIntrvw_LRR_160904.asp).

101. IEA, *World Energy Outlook 2004*.

102. Walter Youngquist, "World Energy Council survey of energy resources: oil shale" (http://www.ecology.com/archived-links /oil-shale/).

103. "Strategic significance of America's oil shale resource, volume 1: assessment of strategic issues", Office of Deputy Assistant Secretary for Petroleum Reserves, Office of Naval Petroleum and Oil Shales Reserves, US Department of Energy, Washington DC, March 2004.

104. "Oil peak – a geologist's view", quoted at note 70.

105. Information in this section thus far from Bob Williams, "Debate grows over US gas supply crisis as harbinger of global gas production peak", *Oil & Gas Journal*, vol. 101, no. 28, 21 July 2003, pp. 20–28; C.J. Campbell, "Industry urged to watch for regular oil production peaks, depletion signals", *Oil & Gas Journal*, vol. 101, no. 27, 14 July 2003, pp. 38–48.

106. Carola Hoyos, "Shortages of easily accessible natural gas mean more interest in technology that allows supplies to be brought from more remote areas", *Financial Times*, 15 August 2003.

107. John McCaughey, "LNG in the US: the gathering storm", *World Energy Review*, June 2004.

108. "Debate grows over US gas supply crisis" and "Industry urged to watch for regular oil production peaks", quoted at note 105.

109. For a more detailed description of hydrates, see my account in *The Carbon War*, quoted at note 10.

110. "Debate grows over US gas supply crisis", quoted at note 105, p. 24.

111. Daniel Yergin, *The Prize: The Epic Quest for Oil, Money and Power*, Simon and Schuster, 1991, pp. 571–574 and 665–666.

112. ASPO's estimate of peak oil has recently been revised down from 2006 to 2005.

113. "The cheap oil era is far from over", *National Geographic News*, 20 May 2004.

114. "Energy: a global overview", quoted at note 88.

115. "Strategic significance of America's oil shale resource", quoted at note 103.

116. Dr W. Stuart McKerrow, a legendary researcher of earth history, who died in 2004.

117. Adrian Michaels in New York and Carola Hoyos in London, "SEC leans to voluntary code on oil reserves", *Financial Times*, 23 June 2004.

118. Kevin Morrison, "Secret and unreliable world of oil statistics", *Financial Times*, 15 December 2004.

119. Colin Campbell and Jean Laherrère, "The end of cheap oil: global production of conventional oil will begin to decline sooner than most people think,

probably within 10 years", *Scientific American*, March 1998, pp. 59–65.

120.   "The end of cheap oil", quoted at note 119.

121.   ASPO newsletters of July and August 2004 and references therein.

122.   Chris Skrebowski, "Oilfield megaprojects", *Petroleum Review*, January 2004.

123.   Chris Skrebowski, "New projects to 2012", *Petroleum Review*, April 2005.

124.   "Energy: a global overview", quoted at note 88.

125.   "Oil peak – a geologist's view", quoted at note 70.

126.   Michael Smith, personal communication, January 2005.

127.   Michael Smith, "The Middle East – miracle or mirage?", Energy Institute conference quoted at note 15.

128.   "The sustainability of higher oil prices", quoted at note 84.

129.   Philip Thornton, "Hidden costs of pipeline meant to safeguard West's oil supply", *Independent*, 26 June 2004. An independent panel set up by BP, the Caspian Development Advisory Panel, which included environmental and human rights activists as well as senior industry and government figures, concluded as follows: "The panel heard concerns that Botas [the company awarded the contract to build the Turkish section] and its contractors might feel pressure to cut corners on environmental, social and technical standards to remain on schedule."

130.   "The end of cheap oil", quoted at note 119.

131.   Christopher Pala, "Kazakh government discourages Caspian exploration", *Petroleum Review*, November 2004.

132.   *Blood and Oil*, quoted at note 23.

133.   "Will China and US have to compete in global search for oil?", Associated Press, 6 October 2004.

134.   Matthew Taylor, "Elderly couple die after gas cut off", *Guardian*, 23 December 2003.

135.   Andrew Warren, Director of the Association for the Conservation of Energy, has long campaigned on this and other crazy aspects of Britain's energy profligacy. See, for example, "A long, cold, wait", *Energy in Buildings & Industry*, March/April 2000.

136.   John Houghton, "Global warming is getting worse – but the message is getting through", *Guardian* Online, 16 August 2004.

137.   Calculation made by Dr Andrew Dlugolecki, formerly a senior executive in General Accident, now a consultant to the United Nations Environment Programme's Finance Initiative.

138.   As described in *The Carbon War*, quoted at note 10.

139.   "Topics: annual review of natural catastrophes 1996", Munich Re Special Publication, 1997.

140.   Reuters report, Geneva, 4 March 2004.

141. *The Carbon War*, quoted at note 10.

142. The Toronto Declaration on Climate Change and Ozone Depletion, where government officials and experts discussed the creation of the Intergovernmental Panel on Climate Change.

143. The Intergovernmental Panel on Climate Change (henceforth IPCC), First, Second and Third Scientific Assessment Reports of 1990, 1995 and 2001, Cambridge University Press.

144. They mix in different proportions in warm air or water, and cold air or water. For further detail, see John Houghton, *Global Warming the Complete briefing*, Cambridge University, 2004, p. 67.

145. IPCC Third Scientific Assessment Report, 2001, quoted at note 143.

146. Chart as used in latest Meteorological Office climate-change presentation (Richard Betts, personal communication, 2004).

147. "Uncertainty, risk and dangerous climate change", Special Publication of the Hadley Centre (Meteorological Office) and the Department of Environment, Food and Rural Affairs, December 2004.

148. With a 90 percent confidence level that it is in this range. "Uncertainty, risk and dangerous climate change", quoted at note 147.

149. Quoted in Fred Pearce, "Kyoto Protocol is just the beginning", *New Scientist*, 10 October 2004.

150. Michael McCarthy, "Countdown to global catastrophe", *Independent*, 24 January 2005. The taskforce comprised the UK's Institute for Public Policy Research in London, the US Center for American Progess in Washington DC and the Australia Institute in Canberra.

151. "Climate change: meeting the challenge together", conference in Berlin, 3 November 2004. Comments made by Sir David King, UK Chief Scientific Adviser, and John Schellnhuber, Research Director, Tyndall Centre (http://www.britischebotschaft.de/statevisit/en/press/steering_paper.pdf).

152. John Schellnhuber, quoted in Fred Pearce, "Act now before it's too late: that was the message from climate researchers attending last week's conference in Exeter, UK", *New Scientist*, 12 February 2004.

153. Most of this amount is below deep water, but significant quantities exist at shallow depths in the Arctic, in danger of being destabilized by the faster-then-average warming there. I discuss methane hydrates in both *The Carbon War*, quoted at note 10, and *Global Warming: The Greenpeace Report*, Oxford University Press, 1990.

154. Summary in "Act now", quoted at note 152.

155. Though not as quickly, or to the absolute extent, depicted in the film *The Day After Tomorrow*.

156. Summary in "Act now", quoted at note 152.

157. The latest Hadley Centre simulation indicates that, by the year 3000, more than half the ice sheet would be gone, raising global sea level up to 4 metres.
158. The paperback edition, quoted at note 10, came out in 2000, with an update.
159. IPCC First Scientific Assessment Report, 1990, quoted at note 143.
160. Article 2 of the Framework Convention on Climate Change reads: "The ultimate objective of this Convention and any related legal instruments that the Conference of the Parties may adopt is to achieve, in accordance with the relevant provisions of the Convention, stabilization of greenhouse gas concentrations in the atmosphere at a level that would prevent dangerous anthropogenic interference with the climate system. Such a level should be achieved within a time-frame sufficient to allow ecosystems to adapt naturally to climate change, to ensure that food production is not threatened and to enable economic development to proceed in a sustainable manner." (http://unfccc.int/files/essential_background/background_publications_htmlpdf/application/pdf/conveng.pdf.)
161. IPCC Second Scientific Assessment Report, 1995, quoted at note 143.
162. IPCC Third Scientific Assessment Report, 2001, quoted at note 143.
163. E. Mielke et al., "Russia and Kyoto: match made in heaven?", Dresdener Kleinwort Wasserstein Research, 2 December 2004.
164. Estimate cited in Merrill Lynch Energy Managers newsletter, 15 November 2004.
165. "World oil and gas 'running out'", CNN, 2 October 2003, quoting Professor Kjell Aleklett.
166. Latest IPCC work summarised by Malte Meinshausen of the Swiss Federal Institute of Technology, "On the risk to overshoot 2°C", in press. See also Bill Hare and Malte Meinshausen, "How much warning are we committed to and how much can be avoided?", Potsdam Institute for Climate Impact Research, PIK Report no 93, October 2004.
167. IPCC figures from the Third Scientific Assessment Report, 2001, quoted at note 143.
168. Donald Aitken, Lynn Billman and Stanley Bull, "The climate stabilization challenge: can renewable energy sources meet the target?", *Renewable Energy World*, November-December 2004.
169. David Dapice, "Coal-to-liquid fuel offers answers to energy woes", *Straits Times*, 19 July 2004.
170. Frank Gray, "Clean coal becomes a reality", *World Energy Review*, August 2004.
171. David Adam, "Oil chief: my fears for planet. Shell boss's 'confession' shocks industry" and "I'm really very worried for the planet': Ron Oxburgh is chipping away at the fossilised thinking that cost Shell its reputation",

Guardian, 17 June 2004.

172.   Sir John Houghton was right. The Reverend Ted Haggard, president of the US National Association of Evangelicals, would say in June 2005: "You don't need scientific research to understand how serious this problem is ... It takes very manipulative people to look the other way ... We do represent 30 million people, and we can mobilize them behind this, and we will, if we have to." ("Evangelical leaders combat global warming", Independent Press news service, March 2005 [no specific date].) The copy refers to the influence Sir John had.

173.   There are interesting comparisons to be made between Richard Lindzen's position in the 1990s and M. King Hubbert's in the 1950s. Brilliant individuals can be proved right, and have been many times in the face of a robust peer-group consensus to the contrary. But there is a big difference between the position taken by lone-voice scientists such as Einstein, Galileo, or whoever, and Richard Lindzen's position. It is that the fate of the planet didn't necessarily hinge on their being right in the event that governments elected to side with them and not the consensus of the day. I confronted Lindzen with this in public debate as long ago as 1991 (see *The Carbon War*, quoted at note 10, pp. 39–41). Thirteen years later there was still not a shred of humility in his presentation, much less recognition of the stakes involved if he were wrong.

174.   *The Prize*, quoted at note 111. Unless otherwise stated, most of the account in this chapter is based on that book and Anthony Sampson's *The Seven Sisters*, Coronet Books, 1975.

175. $100,000 in the money of the day. For the conversion to modern money I used www.cjr.org/tools/inflation/, *Columbia Journalism Review*.

176.   Thomas L. Friedman, "People connected with Exxon reportedly contributed more than $1 million to the Bush campaign", *New York Times*, 1 June 2001; www.opensecrets.org, the website of the Centre for Responsive Politics, cited in "A decade of dirty tricks", Greenpeace briefing, July 2001.

177.   See Michael Moore's *Stupid White Men: And Other Sorry Excuses for the State of the Nation!*, Penguin, 2002, for an analysis of the links between the key figures in George W. Bush's first cabinet and the oil companies.

178.   On 4 April 1953, US CIA Director Allen W. Dulles approved $1 million to be used "in any way that would bring about the fall of Mossadegh". They decided on the spreading of lies and propaganda. CIA operatives posing as Mossadegh supporters threatened Muslim leaders in Iran with savage punishments for opposing their Prime Minister. The plot, codenamed Operation Ajax, stirred the desired political dissent. Anti- and pro-monarchists promptly took to the streets to fight it out. Mossadegh lost,

served three years in prison for treason and lived out the rest of his life under house arrest. In March 2000, the then Secretary of State Madeleine Albright said, "The Eisenhower Administration believed its actions were justified for strategic reasons. But the coup was clearly a setback for Iran's political development and it is easy to see now why many Iranians continue to resent this intervention by America." (James Risen, "Secrets of history: the CIA in Iran, a special report. How a plot convulsed Iran in '53 (and in '79)", *New York Times*, 16 April 2000; www.wikipedia.org.)

179. Quoted in Walter LaFeber, *America, Russia, and the Cold War, 1945–2002*, McGraw-Hill, New York, 2002, p. 292; www.wikipedia.org.

180. *The Prize*, quoted at note 111, p. 614.

181. "The October oil embargo", *Popular Mechanics*, 1996, p. 48, quoted at www.wowessays.com/dbase/ ab1/iev247.

182. *The Prize*, quoted at note 111, p. 625.

183. Lizette Alvarez, "Britain says US planned to seize oil in 1973 crisis", *New York Times*, 2 January 2004.

184. Events in this paragraph to here are as described in Jeffrey Robinson's unauthorized biography of Sheik Yamani, *Yamani: The Inside Story*, Simon and Schuster, 1988, pp. 101–102.

185. *Yamani*, quoted at note 184, p. 102.

186. *Yamani*, quoted at note 184, p. 102.

187. Information in this paragraph from *The Prize*, quoted at note 111, and *The Seven Sisters*, quoted at note 174.

188. Shah Reza Pahlevi was returned as absolute ruler of Iran courtesy of the 1953 CIA-orchestrated coup. In the following twelve years the US poured over $1.2 billion in aid into Iran, almost half of which went to the Iranian army. Between 1950 and 1977, the US supplied Iran with over $20 billion worth of arms, ammunition and assistance under the Military Assistance Program (MAP) and the Foreign Military Sales Program (FMS). This includes $67.4 million to train 11,025 military personnel. Between 1970 and 1978 Iran bought another $18 billion of US arms under the FMS cash sales program. (Iran Virtual Library http://www.irvl.net/.)

189. *The Prize*, quoted at note 111, p. 692.

190. "The sustainability of higher oil prices", quoted at note 84.

191. The war was disastrous for Iraq, stalling economic development, disrupting oil exports and saddling Saddam with $40 billion of debts, including $14 billion loaned by Kuwait (www.wikipedia.org). American companies, with US government backing, supplied Hussein with much of the raw material for Iraq's chemical and biological weapons arsenal. Hussein duly killed upwards of 5,000 civilians in a Kurdish town by gassing them with chemical weapons.

At the time, the incident was played down by the Reagan Administration which claimed that Iran, then the preferred American enemy, was actually responsible. While the Reagan Administration softly chided Hussein's use of chemical weapons in public, the White House secretly sent a certain Donald H. Rumsfeld, then Special Middle East Envoy, to Baghdad in March 1984 to reassure the offended dictator that, really, the US had no problem with his gassing policy, specifically that the US desire to "improve bilateral relations, at a pace of Iraq's choosing", remained "undiminished". (Dana Priest, "Rumsfeld visited Baghdad in 1984 to reassure Iraqis, documents show", *Washington Post*, 19 December 2003.)

192. In late July 1991, just days before invading Kuwait, Saddam summoned US ambassador April Glaspie to meet him. Although Glaspie would later claim she used the meeting to "repeatedly warn Hussein against using force to settle his dispute with Kuwait" (testimony to US Senate, April 1991), her phrasing was obviously not strong enough. Telling Saddam that she came "in the spirit of friendship, not in the spirit of confrontation", Glaspie added: "I have a direct instruction from the President to seek better relations with Iraq", and "… we have no opinion on the Arab-Arab conflicts, like your border disagreement with Kuwait". ("Excerpts from Iraqi document on meeting with US envoy", *New York Times International*, 23 September 1990.)

193. All events described in this section are covered in more detail in *The Carbon War*, quoted at note 10.

194. George Soros, *The Crisis of Global Capitalism: Open Society Endangered*, Little, Brown and Company, 1998.

195. Peter Hetherington and Charlotte Denny, "Oil giants accused of collusion: business may be in cahoots with pickets, says Straw aide", *Guardian*, 13 September 2000.

196. Pamela Najor, "RIP Global Climate Coalition", Bureau of National Affairs, 24 January 2001.

197. John Vidal, "The oil under this wilderness will last the US six months. But soon the drilling will begin", *Guardian*, 18 March 2005.

198. "Fuel economy: going farther on a gallon of gas", Union of Concerned Scientists webpage (http://www.ucsusa.org/clean_vehicles/cars_and_suvs/page.cfm?pageID=222).

199. "The end of the oil age: ways to break the tyranny of oil are coming into view. Governments need to promote them", *Economist*, 23 October 2003. And for an extended account, see Vijay V. Vaitheeswaran, *Power to the People: How the Coming Energy Revolution will Transform an Industry, Change our Lives, and maybe even Save the Planet*, Farrar, Straus and Giroux, 2005 (p/b edition).

200. Craig Unger, *House of Bush, House of Saud: The Secret Relationship Between the World's Two Most Powerful Dynasties*, Scribner, 2004 (p/b edition).

201. Dominic White, "Shell drops 'bombshell' on reserves", *Daily Telegraph*, 10 January 2004.

202. Paul Brown and Mark Oliver, "Top scientist attacks US over global warming", *Guardian*, 9 January 2004.

203. David Stipp, "Climate collapse: the Pentagon's weather nightmare. The climate could change radically, and fast. That would be the mother of all national security issues", *Fortune*, 26 January 2004.

204. "Major institutional investors back global warming resolutions", note in my diary, 27 February 2004.

205. Speech by Lord Browne, Group Chief Executive, BP plc, at National Press Club, Washington DC, 23 March 2004.

206. Norm Cohen and Clay Harris, "Shell shock: human failings and hyperbolic e-mails", *Financial Times*, 20 April 2004; Terry Macalister, "Shell admits it misled its investors", *Guardian*, 20 April 2004.

207. Julian Borger in Washington and Luke Harding in Baghdad, "Rumsfeld: I won't quit. Pentagon chief given Senate grilling. Abuse of Iraqis rife, says Red Cross", *Guardian*, 8 May 2004.

208. Terry Macalister, "BP ready to quit [Iraq] in blow to rebuilding hopes", *Guardian*, 29 April 2004.

209. Terry Macalister, "Shell hopes Iraq can plug oil leak", *Guardian*, 4 May 2004.

210. "Coping with sky high oil prices", *Business Week*, 30 August 2004.

211. James Boxell, "World oil reserves up 10%, says BP", *Financial Times*, 16 June 2004.

212. "Experts warn business world is running out of oil surplus", *Voice of America* news, 16 June 2004.

213. Bruce Bartlett, "Doom-and-gloomers say we're near tapped-out of oil. Rubbish", ENATRES, 9 June 2004.

214. Larry Elliott, "Even OPEC cannot stem this surge: supply and demand issues are forcing up oil prices", *Guardian*, 2 June 2004.

215. Terry Macalister, "Once seen as an alarmist fear, an attack on key Saudi oil terminal could destabilise west", *Guardian*, 3 June 2004.

216. Jonathan Steele and Terry Macalister, "Vital oil exports halted after sabotage. Political handover dealt blow as insurgents wreck pipelines and assassinate top industry executive", *Guardian*, 17 June 2004.

217. "Seeking a unified approach – valuing and booking oil reserves", London Energy Group seminar, 5 July 2004. The quote is from William Prast, an oil and gas consultant to the Financial Services Authority. He also said: "Uncertainty is paramount, so much so that most petroleum engineers should

have been awarded degrees in arts not sciences."

218. Adam Sieminski and Jay Saunders, "Hubbert's pique", Deutsche Bank Global Energy Wire, 9 June 2004.

219. Various articles in *Financial Times*, 29 June 2004.

220. Terry Macalister, "Shell's shame: FSA spells out abuse", *Guardian*, 25 August 2004.

221. Michael Harrison, "Shell pays £83 fine to settle scandal over reserves", *Independent*, 30 July 2004.

222. "Can oil majors keep up with worldwide demand?", *Houston Chronicle*, 30 July 2004.

223. Sadad al-Husseini, "Why higher oil prices are inevitable this year, rest of decade", *Oil & Gas Journal*, 2 August 2004.

224. Ashley Seagar, "Petrol prices unlikely to fall again as OPEC warns on supply", *Guardian*, 4 August 2004.

225. Ashley Seagar, "Oil threat to world economy", *Guardian*, 5 August 2004.

226. Carolynne Wheeler and Ashley Seager, "Oil prices at record high after Kremlin u-turn on Yukos", *Guardian*, 6 August 2004.

227. Oliver Morgan, Sarah Ryle and Edward Helmore, "Oil and US jitters prompt fear of global slowdown", *Observer*, 8 August 2004.

228. "The devil's curse", *Guardian* Leader, 7 August 2004.

229. Larry Elliott, "Iraq sabotage fear deepens oil crisis", *Guardian*, 10 August 2004.

230. "Deutsche Bank warns oil price may hit $100", quoted in September 2004 ASPO newsletter.

231. Carola Hoyos, "Call to let in oil majors to boost output", *Financial Times*, 13 September 2004.

232. James Boxell and Carola Hoyos, "BP chiefs claims at odds with rivals", *Financial Times*, 17 September 2004.

233. James Boxell and Carola Hoyos, "Oil majors well placed to meet growth in demand", *Financial Times*, 16 September 2004.

234. Kevin Morrison, "Exxon chief hits at energy debate", *Financial Times*, 17 September 2004.

235. Kevin Morrison, Javier Blas and James Boxell, "OPEC raises quotas to deflect criticism", *Financial Times*, 16 September 2004.

236. William Keegan, "Pouring oil on troubled economists", *Observer*, 10 October 2004.

237. Christopher Brown-Humes, "Oil prices hit records after fears on output", *Financial Times*, 12 October 2004. This is the price of Brent crude.

238. Terry Macalister, "Oil hits record $54 a barrel", *Guardian*, 13 October 2004.

239. "Crude oil surges to new high", Associated Press, 25 October 2004.

240. Doug Cameron, Kevin Morrison and Javier Blas, "US rejects calls by OPEC chief to free oil reserves", *Financial Times*, 27 October 2004.

241. Mark Milner, "Browne calms oil supply fears. Surging prices push BP's profits up 43% to $4 billion", *Guardian*, 27 October 2004.

242. "The next scandal?", *Economist*, 13 November 2004.

243. "Big oil's biggest monster", *Economist*, 8 January 2005.

244. "Expert: Saudi oil may have peaked", aljazeera.net, 20 February 2005.

245. "Fuel's gold: heading for new peaks", *Economist*, 12 March 2005.

246. "White House: oil costs due in part to lack of legislation", Dow Jones newswires, 17 March 2005.

247. Javier Blas and Kevin Morrison, "IEA to call for an emergency oil plan", *Financial Times*, 2 April 2005.

248. David Lazarus, "ChevronTexaco CEO banking on peak oil situation", *San Francisco Chronicle*, 8 April 2005.

249. "Bank says Saudi's top field in decline", aljazeera.net, 12 April 2005.

250. "Energy needs, choices, and possibilities: scenarios to 2050", Shell special publication, 2001. In this scenario, only in Europe would energy-efficiency measures be needed in order for renewables to hit the target.

251. I first heard this from Roger Booth, when he was Head of Renewables at Shell, in the mid-1990s. Roger subsequently joined me as a director of solarcentury in 1999, and he remains a key adviser to the company to this day.

252. Global annual electricity use at ten terawatt years = 87,600,000,000 kilowatt hours. The number of kilowatt hours per kilowatt of peak power of solar PV in a region as sunny as the Sahara would be 2,000. Therefore the kilowatts of peak power would be 43,800,000,000. 10 square metres delivers 1kWp, which is 438 billion square metres, equal to a square 600 by 600 kilometres.

253. "Solar energy: brilliantly simple", BP pamphlet, available on UK petrol forecourts.

254. US Department of Energy figures quoted in Lester Brown, *Eco-economy: Building an Economy for the Earth*, Earth Policy Institute, 2001.

255. *Eco-economy*, quoted at note 254.

256. For detailed argument see Amory Lovins, E. Kyle Datta, Odd-Even Bustnes, Jonathan G. Koomey and Nathan J. Glasgow, *Winning the Oil Endgame: Innovation for Profit, Jobs, and Security*, Earthscan, 2004, p. 240.

257. /258 "Renewable energy: practicalities", Report of the House of Lords Science and Technology Committee, HL Paper 126-I, 2004.

258. "You want the confidence to invest in renewable energy", Department of Trade and Industry information booklet, 2004.

259. "You want the confidence", quoted at note 258. They say 20,000 homes, but use the average home consumption, which is of course energy inefficient.

260. "A microgeneration manifesto", Green Alliance, September 2004. The tens of thousands of small sites could make 500 megawatts to 2 gigawatts.

261. Anaerobic digestion is the breaking down of carbohydrates by bacteria in the absence of oxygen. Gasification entails heating wood chips on a "fluidized bed": a controlled flow of air of steam. The result is a combustible mixture of gases, including carbon monoxide, hydrogen and methane. Pyrolysis entails the heating of organic waste at high pressure, in the absence of air. The result is a high-quality oil that can be used to fuel a power plant. These techniques are more efficient than traditional combustion because the gas, mixed with air, burns at a higher temperature, so driving the turbine more efficiently.

262. Patricia Thornley, biomass expert on the UK Government's Renewables Advisory Board, personal communication.

263. Rather less electricity than in a conventional power plant.

264. See *Power to the People*, quoted at note 199 above.

265. John Gartner, "Automakers give biodiesel a boost", *Wired*, 23 September 2004.

266. Information in this paragraph from "Field of dreams: is turning crops into fuels and chemicals the next big thing?", *Economist*, 7 April 2004.

267. Many fuel cells are of a "proton exchange membrane" (PEM) type. In these, a solid polymer is sandwiched between two electrodes (an anode and a cathode). A platinum catalyst stimulates hydrogen to give up an electron at the cathode. The proton travels through the membrane to the cathode, but the electron has to move all the way round it via an electrical circuit that can be tapped usefully. At the cathode, the electrons and protons recombine making clean water. For further details see Peter Hoffman, *Tomorrow's Energy: Hydrogen, Fuel Cells and the Prospects for a Cleaner Planet*, The MIT Press, 2002.

268. In steam reforming, a hydrocarbon is first vaporized in a combustion chamber, and then flows into a steam reformer chamber, where a catalyst breaks apart the gases and water vapour. For a diagram and explanation, see Matthew L. Wald, "Questions about a hydrogen economy", *Scientific American*, May 2004, pp. 66–71.

269. Electrolysis involves passing an electric current through water to split the molecules. Oxygen atoms are attracted to the anode (positive terminal) and hydrogen atoms are attracted to the cathode (negative terminal).

270. "Questions about a hydrogen economy", quoted at note 268.

271. Craig Simons, "The high road: if China steers its auto industry toward hybrids and perhaps hydrogen cars, the world may have no choice but to follow", *Newsweek*, 6–13 September 2004.

272. *Winning the Oil Endgame*, quoted at note 256.

273. The Condé Nast Building in Manhattan is a prominent American example. In the UK, Woking Borough Council's innovative low-carbon programme includes a large fuel cell at a sports complex.

274. Publicity pamphlets of Ovonic Batteries, a subsidiary of Energy Conversion Devices.

275. "Hybrid future: Sky high prices are driving a quest for efficient energy", *Newsweek*, 8 September 2004.

276. *Winning the Oil Endgame*, quoted at note 256.

277. "The sustainability of higher oil prices", quoted at note 84.

278. Nelson Schwartz, "Poor little rich company: on the back of $55 oil, ExxonMobil has become one of the world's richest companies. And that's the problem", *Fortune*, 18 April 2005.

279. "Hubbert's Curve", quoted at note 60, pp. 56–61 (interview with M. King Hubbert).

280. "We can move towards a near-zero carbon future, says Margaret Beckett", Department for Environment, Food and Rural Affairs news release, 4 February 2003. The report, "Assessment of technological options to address climate change – a report to the Prime Minister's Strategy Unit", can be accessed from the Cabinet Office's Strategy website at http://www.strategy.gov.uk.

281. Allan Jones MBE, "Woking: local sustainable energy community", presentation to Low Carbon Thames Gateway conference, Barking, 16 June 2004.

282. *Winning the Oil Endgame*, quoted at note 256.

283. Robert Monks, *The New Global Investors*, Capstone Press, 2001.

284. Clayton Christensen, *The Innovator's Dilemma: When New Technologies Cause Great Firms to Fail*, Harvard Business School Press, 1997.

285. I presented a version of this optimistic analysis in a written and oral paper, "The future of energy supply: civilization at risk", to the Institute of Petroleum conference, "Fuelling the world economy: future risks and opportunities", in London, 17 February 2003. The oral presentation was given to a thin and hostile audience. The paper was later published in the first edition of *New Academy Review*.

286. Alex Kirby, "When the last oil well runs dry", BBC News Online, 16 April; quote by Simmons from May 2003 Paris meeting of ASPO.

287. John McGaughey, "Energy reserves: apocalypse now, tomorrow or never?", *World Energy Review*, August 2004.

288. By hydrogen fuel plants I mean those using fossil-fuel feedstock, not small-scale hydrogen-from-renewables plants.

289. Gordon MacKerron, "Nuclear power and the characteristics of 'ordinariness'

– the case of UK energy policy", *Energy Policy*, vol. 32, pp. 1957–1965.

290. Department of Trade and Industry, *Our Energy Future – Creating a Low Carbon Economy*, The Stationary Office, 2003, pp. 61–62.

291. As I heard live in presentations to government by energy-industry executives during the consultation for the 2003 Energy White Paper.

292. I knew Ted Taylor well in the late 1980s, when we both attended meetings of the scientists' arms-control lobby group, Pugwash. He helped me greatly at the time I set up the technical think-tank VERTIC (the Verification Technology Information Centre), with money from Quaker foundations.

293. Rob Edwards, "The nightmare scenario: what would happen if a passenger jet ploughed into a nuclear plant?", *New Scientist*, 13 October 2001, pp. 10–12. Peter Bunyard, "The plane truth: Terrorists don't need nuclear weapons when there are ready-made atomic bombs awaiting detonation by a hijacked aircraft loaded with fuel", *The Ecologist*, vol. 31, no. 9, November 2001.

294. "The PBMRL: 'old wine in a new bottle'", Nuclear Information and Resource Service (http://www.nirs.org/factsheets/PBMRFactSheet.htm); website accessed 5 November 2004.

295. Svetlana Alexievich, "Land of the dead: remembering the Chernobyl disaster by those who survived it", *Guardian*, 25 April 2005.

296. "A tale of two planets", quoted at note 51.

297. John Carey, "Global warming: why business is taking it so seriously", *Business Week*, 30 August 2004.

298. "Fortune magazine backs renewable energy", 16 August 2004, www.solaraccess.com.

299. New Energies Invest AG (www.neweenergiesinvest.ch).

300. *Photon*, March 2004. The oil index quoted is the OXI.

301. Michael Rogol, Shintaro Doi and Anthony Wilkinson, "Sun screen: Investment opportunities in solar power", Crédit Lyonnais Securities Asia report, July 2004.

302. "CO_2 capture and storage", website maintained by IEA Greenhouse Gas R&D Programme, http://www.co2sequestration.info/research_programmes.htm.

303. Mark Clayton, "America's new coal rush: utilities' dramatic push to build new plants would boost energy security but hurt the environment", *Christian Science Monitor*, 26 February 2004.

304. Karen Armstrong, *The Battle for God*, HarperCollins, 2001 (p/b edition).

305. Michael Crichton, *State of Fear*, HarperCollins, 2004. And see my review of the book in "Making a myth of climate change", *New Scientist*, issue no. 2489, 2 March 2005.

306. Information in this paragraph from "Carbon sequestration – technology

roadmap and program plan, 2004", US Department of Energy and National Energy Technology Laboratory, April 2004 (http://www.fe.doe.gov/ programs/sequestration/publications/ programplans/2004/ Sequestration Roadmap4-29–04.pdf).

307. Vanessa Houlder, "The case for carbon capture and storage", *Financial Times*, 23 January 2004.

308. Nicola Jones, "Bubbling under", *New Scientist* Breaking News, 20 June 2001.

309. Information in this paragraph from Maggie McKee, "Seas absorb half of carbon dioxide pollution", *New Scientist*, 15 July 2004.

310. IPCC figures from the *Third Scientific Assessment Report*, 2001, quoted at note 143.

311. James Randerson, "Forest experiment questions greenhouse gas strategy", *New Scientist* Breaking News, 15 April 2002.

312. Nicola Jones, "A risk too far: dumping CO_2 in the oceans could be a disaster, yet it's still legal", *New Scientist*, 20 October 2001.

313. Fred Pearce, "Scientists use creativity to fight global warming", Global Newspaper Company, 20 January 2004.

314. "Approaching global crisis threatens to reverse human development", *Environment Times*, 22 June 2004.

315. "Approaching global crisis", quoted at note 314.

316. Mark Moody-Stuart et al., G8 Renewable Energy Task Force, Final Report, July 2001.

317. *Power to the People*, quoted at note 199, cites 92–98 percent repayment rates, p. 313.

318. Joanna Chung, "States file suit to cut levels of CO_2", *Financial Times*, 22 July 2004.

319. Myles Allen and Richard Lord, "The blame game: who will pay for the damaging consequences of climate change?", *Nature*, vol. 432, pp. 551–552, 2 December 2004.

320. Peter Roderick, "Damage litigation", *Guardian*, 15 December 2004.

321. D.G. Cogan, "Corporate governance and climate change", Ceres, Boston, 2003.

322. "Energy in Focus", quoted at note 19.

323. ASPO newsletter, August 2004.

324. "The sustainability of higher oil prices", quoted at note 84.

325. Stephan Schmidheiny, Federico Zorraquin and the World Business Council for Sustainable Development, *Financing Change: The Financial Community, Eco-efficiency, and Sustainable Development*, The MIT Press, 1996.

326. Jeremy Leggett, ed., *Climate Change and the Financial Sector: The Emerging Threat, the Solar Solution*, Gerling Akademie Verlag, 1996.

327. "The investor guide to climate risk", Ceres, July 2004, written by the Investor Responsibility Research Center, an investor advisory firm. The guide was commissioned on behalf of the Investor Network on Climate Risk (INCR), a new alliance of institutional investors dedicated to promoting better understanding of the risks of climate change among institutional investors.

328. Note from my diary, March 2001. Other thinkers in the communications arena talk of the value of trust being so important for businesses in the years to come that corporations will have to behave increasingly like campaigning organisations.

329. Malcolm Gladwell, *The Tipping Point: How Little Things Can Make a Big Difference*, Back Bay Books, 2002 (p/b edition).

330. Stephen Leeb and Donna Leeb, *The Oil Factor: Protect Yourself – and Profit – from the Coming Energy Crisis*, Warner Business Books, 2004.

331. See *The Carbon War*, quoted at note 10.

332. Saudi Arabia (1), Russia (2), Libya (4), China (6), United Arab Emirates (10), Kuwait (12), Nigeria (13) and Iran (14). Source: "Energy in Focus", quoted at note 19.

333. The US Air Force attacked al-Jazeera's Baghdad studio with missiles during the second Gulf War. Al-Jazeera had been reporting from the streets to 40 million Arab-speaking viewers, showing pictures of maimed Iraqi civilians hit by bombing at the time the Americans wanted everyone to believe they were taking out only precisely defined targets in Baghdad. Donald Rumsfeld said of al-Jazeera at the time: "Ultimately people are caught lying and they lose their credibility." Indeed. One day we may know who in the US high command ordered the murderous strike, which was actually filmed by an al-Jazeera cameraman on the roof. *The Control Room*, a documentary film transmitted on BBC4 on 21 August 2004, shows the footage. The US military claimed the aircraft had come under fire from the ground.

334. See Anthony Giddens, *Runaway World: How Globalisation is Shaping our Lives*, Profile Books, 1999, p. 102.

335. Tim Radford, "Oceans of evidence for global warming", *Guardian*, 19 February 2005. A report from the American Association for the Advancement of Science, not published in the academic press at the time of writing.

336. See the oral account of C. Wright Patman, long-serving Texan Congressman, in *Hard Times*, quoted at note 30, p. 285. "At the time, they [the moneyed interests] thought they'd get a dictatorship here. General Smedley Butler was picked out to be the leader. He was gonna be their man on the white horse. They were gonna close in and take this country over. And they come darn near doing it."

337. *Runaway World*, quoted at note 334, p. 102.

338. Colin Hines, *Localization: A Global Manifesto*, Earthscan, 2000, p. 209.
339. *Eco-economy*, quoted at note 254.

Index

Numbers in italics indicate Figures.

unconventional gas
gasification 204, 290n
gasoline queues 144, 171
GCC *see* Global Climate Coalition
Gelsenkirchen solar photovoltaic
    manufacturing plant 134
General Electric 218
General Motors 22, 140, 148, 182, 206,
    208, 210, 247–8
Geological Museum, London 273n
geologists 25
Germany, and PV 233
Ghawar field, Saudi Arabia 36, 49, 50,
    58, 60, 67, 91, 193
Gheit, Fadel 187
Gladwell, Malcolm: *The Tipping Point*
    250–52
Glaspie, April 286n
global average temperature 114–15, *115*
Global Climate Coalition (GCC) 157,
    159, 160, 161, 181, 245, 256
global instrumental observations 114
global warming 12, 13, 14, 17, 96,
    103–134, 274n
    conflating oil depletion and global
        warming 126–34
        the carbon arithmetic 127–9
        the spectre of a rush to coal
            129–30
    feedbacks 117–21
    first warning, and emergence of
        Carbon Club 154–6
    how much warming and how much
        danger? 112–17
        the danger threshold 115–17
    international effort to respond to the
        danger so far 122–6
        Convention on Climate Change
            (1992) 122–3
        Kyoto Protocol (1997) 123–4
        since Kyoto 124–6
    as a weapon of mass destruction
        104–112
        twelve reasons for concern
            110–112
        wasting economies and
            ecosystems alike 109–110

Goebbels, Joseph 133
Goering, Hermann 139
Goldman Sachs 67–8, 92–3, 152–3, 168,
    213, 246, 248
Gore, Al 132
governments 242–4
    creation of IPCC 122
    as late toppers 24–5
Great Awakening 270, 271
Great Crash (1929) 229, 265
Great Depression (1930s) 29
Great Underground Cook-Ups 4–6, 12,
    16, 36
Greater Burgan field, Kuwait 58
greenhouse effect 32
    carbon dioxide as a greenhouse gas
        11, 113, 118
    and Climate Convention 123
    CNN coverage 108
    domestic appliances 175
    emissions from homes 97
    and global average temperature 122
    Kyoto Protocol 106, 124
    leaders discuss build-up 13
    methane as a greenhouse gas 118
    and nuclear power 226
    potential global insurance crash 109
    and rate of warming 111
    and tar sands 70
Greenland 111, 113, 118, 119–20
Greenpeace International 126, 181,
    224, 234
Greenpeace UK 133
Guardian 172, 173
Gulf 135–6, 147
Gulf of Mexico 50, 54, 66
Gulf Stream 111, 119
gypsum 37, 38

Hadley Centre simulations 283n
Haggard, Reverend Ted 284n
Halliburton 41–2, 181
Hanifa field, Saudi Arabia 49
Hardman, Richard 89
Harper, Francis 61, 62, 66, 71–2, 73, 90
Hawaii 116
Hayward, Tony 176